TRANSPARENCY DIRECTORY
WITH TEACHER'S NOTES

HOLT, RINEHART AND WINSTON

Harcourt Brace & Company

Austin • New York • Orlando • Atlanta • San Francisco • Boston • Dallas • Toronto • London

HOLT BioSources ™

TRANSPARENCY DIRECTORY WITH TEACHER'S NOTES

Acknowledgments

Editorial Development
Ann Tarleton

Copyediting
Steve Oelenberger
Denise Haney
Amy C. Daniewicz

Prepress
Susan Mussey

Manufacturing
Susan Mussey

Design Development and Page Production
Morgan-Cain & Associates

Cover
Design—Morgan-Cain & Associates
Photography—Sam Dudgeon

ILLUSTRATION CREDITS
Abbreviation Code

BD=Barbe Dekeyser; CF=Chris Forsey; DA=David Ashby; DB=David Beck; DD=David DeGasperis; DF=David Fischer; FP=Felipe Passalacqua; GA=Graham Allen; HH=Henry Hill; JE=John Edwards & Associates; JS=John Seymore; KC=Kip Carter; MCA=Morgan Cain & Associates; MC=Michael Courtney; MHM=Mountain High Map resources © 1993 Digital Wisdom, Inc. MM=Michael Maydak; MW=Michael Woods; PG=Pond & Giles; SK=Steven Kirk; SP=Stephanie Pershing; SW=Sarah Woodward; TH=Tonya Hines

1 Lynne Prentice/LPI; **1a** MCA; **2** MCA; **2a** MCA; **3** MCA; **3a** MCA; **4** Robert Margulies/MMA; **4a** MCA; **5** MCA; **5a** MCA; **6** MCA; **6a** MCA; **7** MCA; **7a** MCA; **8** Robert Margulies/MMA; **8a** MCA; **9** Catherine Twomey/Craven Design Studios Inc.; Robert Margulies/MMA; **9a** MCA; **10** Cynthia Turner Alexander; **10a** MCA; **11** Robert Margulies/MMA; **11a** MCA; **12** MCA; **13** SW; **14** (t) PG; (b) MCA; **15** MCA; **16** MCA; **17** MCA; **18** (tr) © Walker England/PR; (tc) © 1977 Walker England/PR; MCA; **19** MCA; **20** MCA; **21** MCA; **22** MCA; **23** MCA; **24** MCA; **25** Karen Kluglein/MA; **25a** PG; (graphics) MCA; **26** PG; **26a** MCA; **27** MCA; **27a** MCA; **28** MCA; **28a** MCA; **29** Karen Kluglein/MA; **29a** T Narashima; **30** Karen Kluglein/MA; **31** PG; (graphics) MCA; **31a** MCA; **32** PG; (graphics) MCA; **32a** MCA; **33** MCA; **34** MCA; **35** MCA; **36** MCA; **37** SP/Foca; **38** MCA; **39** MCA; **40** MCA; **41** B Hansen/NSI; **42** PG; **42a** MCA; **43** DF; **43a** DA; (graphics) MCA; **44** (map) Lazlo Kubinyi/Rep. Gerald & Cullen Rapp, Inc.; **44a** SW; **45** MCA; **45a** MCA; **46** Yemi; **48** B Hansen/NSI; **49** (l) DA; (r) CF; **50** DA; **51** E Alexander; **52** Daniel Kirk/BA; **54** (jaws) Dana Geraths; (bl,br) SP/Foca; **55** Mark Hallet/MHI; **57** David Beck/RMF; **58** (fern,flowers,seeds) SP/Foca; (moss) © 1989 K.G. Vock/Okapia/PR; (tree) © 1991 Tom Bean/AS; (system) © P. Dayanandan/PR; **59** DA, DD/JE, PG; **60** Susan Johnston Carlson/MT; **61** MCA; **62** Mark Hallet/MHI; **63** PG; **63a** MCA; **64** MCA/MHM; **64a** MCA; **65** MCA; **65a** MCA/MHM; **66** (birds) Daniel Kirk; (map) MCA/MHM; **67** Howard S. Friedman; **68** DB; **69** DB; **70** DF; **71** B Hansen/NSI; **72** (plant,jeep,bkgd) Claire Booth/CBMI; (deer,well) Lynne Prentice/LPI; **73** (bkgd, plants,soil bacteria) Carlyn Iverson; (molecules) Claire Booth/CBMI; (roots,deer) Lynne Prentice/LPI; **76** MCA; **77** SP/JE; **78** (tress) MCA; MW; **79** DB; **80** (t) © Francis, Donna Caldwell/Affordable Photo Stock; (c,b) © Gary Braasch/WCA; **81** (c) Randall Zwingler/LM; (l) © 1989 D.A. Glawe, Univ. of Ill./BPS (tr) © 1991 Ray Pfortner PA; (br) © Master Communications; **82** Carlyn Inverson; **83** (l) Randall Zwinger/LM; (r) Alan Blank/BC; **84** David Beck/RMF; **85** Carlyn Iverson; **86** SP/Foca; **87** Wendy Smith-Griswald/MT; **88** (graphics) MCA; **88a** MCA; **90** MCA; **91** (bkgd) John Coulter/VU; (c) Robert Margulies/MMA; (b) Foca; **92** (b) © Ed Reschke/PA; (t) © John D. Cunningham/VU; (c) © SPL/PR; **93** MCA; **94** MCA; **95** (c) John W. Karapelou/KBI; (b) Foca; **96** MCA; **97** (t) A.J. Olsen © 1986 The Scripps Research Institute (b) John Karapelou/KBI; **98** (l) © A.M. Siegelman/VU; (r) © M. Abbey/VU; **99** MCA; **100** (c) Laurie O'Keefe; (bkgd) Michael P. Gadomski/PR; (b) Foca; **101** FP; **103** HH/JE; **104** HH/JE; **107** MCA; **108** HH/JE; **109** HH/JE; **110** HH/JE; **111** (mushroom) FP; (graphics) MCA; **112a** Wendy Smith-Griswald; **113** MCA/MHM; **113a** PG; **114** Joe LeMonnier; **114a** PG; **115** MCA; **115a** PG; **116** PG; **116a** (t) MCA; (b) PG; **117** PG; **117a** PG; **118** PG; **119** MCA; **119a** MCA **120** MCA; **120a** MCA; **121** PG; **122** (tl) Dr. Jeremy Burgess/SPL/PR; (cr) SP/Foca; (tr) C. Gerald Van Dyke/VU; (cl) Robert Margulies/MMA; (bl,br) E. Alexander/AT; **123** (cl,cr) MCA; **124** (tc) MCA; **125** MCA; **126** MCA; **127** MCA; **128** PG; **129** MCA; **130** (t) MCA; (b) DA; **131** PG; **132** (tl,bc) Deborah Daugherty/HS; (bl) SP/Foca; **133** PG; **134** Laurie O'Keefe; **134a** HH/JE; (graphics) MCA; **135** (fluke) Claire Booth/CBMI; (wasp) Deborah Daugherty/Laurie O'Keefe/HS; (sponge, worm, nematode,snail) Laurie O'Keefe/HS; (hydra,lancelet,seastar) T. McDermott/HS; **135a** FP; **136** (flea) John D. Cunningham/VU; SP/Foca; **136a** JS/JE; **137** (chrysalis,emerging) © Dan Kline/VU; (egg) © Dick Poe/VU; (adult) © E.R. Degginger/AA; (larva) © William J. Weber/VU; **137a** DA, PG; (graphics) MCA; **138** FP; **138a** MCA; **139** FP; **139a** MCA; **140** MCA; **140a** SW; **141** Cecile Duray-Bito; **141a** (dinosaurs) CF; (animals) PG; (graphics) MCA; **142** MCA; **142a** MCA/MHM; **143** C. Turner/AT; **143a** PG; **144** HH/JE; **145** HH/JE; **146** JE/DF; **147** HH/JE; **148** SW; **149** DD/JE; **150** DD/JE; **151** BD; **152** MCA; **153** (t) T. McDermott/Michael Kress-Russick/TMMK; **154** Lynne Prentice/LPI; **155** C. Turner/AT; **156** (t) C. Turner/AT; (b) Michael Abbey/PR; **157** (t) Claire Booth/CBMI; **158** Laurie O'Keefe; **159** Laurie O'Keefe; **160** Laurie O'Keefe; **161** Deborah Daugherty/Laurie O'Keefe/HS; **162** Andrew Grivas; **163** T. McDermott/Michael Kress-Russick/TMMK; **164** Laurie O'Keefe; **165** PG; **166** SP/Foca; **167** (tr,tl) (limbs) Peg Gerrity/HS; **168** David Beck/RMF; **169** (b) SP/Foca; (#3) © 1973 Stephen Dalton/PR; (#2) © 1982 Stephen Dalton/PR; (t) © Hans Pfletschinger/PA; **171** Kip Carter/KC; **172** (frog) SP/Foca; (hearts) Peg Gerrity/HS; (fish) © Tom McHugh/Steinhart Aquarium/PR; **173** Pond & Giles/PS; **175** Lynne Prentice/LPI; **176** Lynne Prentice/Craig Tyler/LPI; **177** Lynne Prentice/Craig Tyler/LPI; **178** Lynne Prentice/LPI; **179** Pat Ortega; **180** KC; **181** KC; **182** KC; **183** (l) MM; (r) DF; **184** MC; **184a** MC; **185** SP/Foca; **185a** MCA; **186** MC; **186a** MC; **187** FP; **187a** MC; **188** Robert Margulies/MMA; **188a** MCA; **189** MC; **190** MC; **190a** MCA; **191** TH; **192** MCA; **193** MC; **193a** MCA; **194** Kevin A. Somerville/CBMI; **194a** FP; **195** MCA; **195a** FP; **196** MC; **197** MC; **197a** MC; (graphics) MCA; **198** (joints) SP/Foca; (schematics) Peg Gerrity/HS; **198a** FP; MCA; **199a** MCA; **200** MC; **200a** MCA; **201** (cl, bl) SP/Foca; (t,br,c) T. McDermott/Michael Kress-Russick/TMMK; **202** MC; **202a** MCA; **203** (t,r) SP/Foca; (bl) John W. Karapelou/KBI; **203a** MCA; **204** Peg Gerrity/HS; **205** TH; **206** Walter Stuart/RWS; **207** Walter Stuart/RWS; **208** Kevin Somerville/MIS; **209** TH; **210** MCA; **211** Walter Stuart/RWS; **212** C. Turner/AT; **213** MCA; **214** MCA; **215** MC; **216** MC; **217** MCA; **218** (l) DF; (lc) MM; (r) MCA; **219** (b) Kevin Somerville/MIS; (tl) T. McDermott/Michael Kress-Russick/TMMK; **220** Yemi; **221** FP; **222** Courtesy USDA; **223** Robert Margulies/MMA; **224** Robert Margulies/MMA; **225** Peg Gerrity/HS; **226** MCA; **227** MC; **228** SW; **229** FP; MCA; **230** MCA; **231** (thyroid) TH; (graphics) MCA; **232** (tc) DF; (r) MC; **233** MCA; **234** MCA; **235** Claire Booth/CBMI; **236** (cells) Walter Stuart/RWS; (plasma) Robert Margulies/MMA; **237** MCA; **238** Claire Booth/CBMI.

PHOTOGRAPHY CREDITS
Abbreviation Code

AA=Animals, Animals; BPA=Biophoto Associates; BPS=BPS & Biological Photo Service; CM=Custom Medical Stock Photo; ES=Earth Scenes; GH=Grant Heilman Photography; JS/FS=Jeff Smith/FotoSmith; NASC=The National Audubon Society Collection; PA=Peter Arnold, Inc.; PH=Phototake NYC; PR=Photo Researchers; RLM=Robert & Linda Mitchell; SP/FOCA=Sergio Purtell/Foca Co., NY, NY; SPL=Science Photo Library; SS=Science Source; TF=Tim Fuller; TSI=Tony Stone Images; VU=Visuals Unlimited

TITLE PAGE: Sam Dudgeon; **12a** (l) CNRI/SPL/PR; (r) M.I. Walker/PR; **24** © RLM; **53** (gibbons) E.R. Degginger/AA; (orangutan) Christopher Arnesen/TS; (gorilla) Pat Crowe/AA; (bonobo) Michael Dick/AA; (chimpanzee,humans) SP/Foca; **63a** (l) © Manfref Danegger/PA; (r) © RLM; **88** (tl) David Muench/TSI; (tr) © Corale Brierly/VU; (bl) © Prof. P. Motta/Dept. of Anatomy/University "La Sapienza", Rome/SS/PR; (br) © 1977 David Scharf; **89** (l) © David M. Phillips/VU; (c) © David Scharf/PR; (r) © R. Kessel-G. Shih/VU; **93** (t) © Hans Gelderblom/VU; (b) © M. Wurtz/Biozentrum, Univ. of Basel/SS/PR; **94** (tl) Runk/Schoenberger/ GH; (tr) © Oscar Bradfute/PA; (b) © K.G. Murti/VU; **112** TM; **118a** (lupine) © William H. Mullins/PR; (clover) © Jack Wilburn/ES; (buttercup, ivy) © Patti Murray/ES; **123** (tl,tr) Runk Schoenberger/GH; (br) © John D. Cunningham/VU; **124** (fibrous root, tap root) Runk Schoenberger/GH; (cross-section) © R.F. Evert; (prop roots) Jane Grushow/GH; (aerial roots) © RLM; **125** (t) © E.R. Degginger/ES; **126** (l) © E.R. Degginger/ES; **138a** © Charles V. Angelo/NASC/PR; **144a** © JS/FS; **170** SP/Foca; **180** © SP/Foca; (t) Pat Ortega; (b) © 1991 John Cancalosi; (crocodylia) © 1991 M.H. Sharp/PR; **182** © SP/FOCA; **189** (l) © BPA/SS/PR; (tr) © Ed Reschke/PA; (br) © SIU/PA; **191** (tr) © David M. Phillips/VU; (cr) © Don Fawcett/SS/PR; (bl) © Ed Reschke/PA; (bc) © R. Calentine/VU; (br) © Bruce Iverson/VU; **193** (l) © Dr. R. Kessel/PA; (r) © Manfred Kage/ PA; (br) © Ed Reschke/PA; **196** (t) © Frederick C. Skvara. M.D.; (b) © Ed Reschke/PA; **199** SPL/CM; **210** © Martin M. Rotker; **232** © David M. Phillips/VU

Printed in the United States of America

ISBN 0-03-050703-0

1 2 3 4 5 6   294   99 98 97 96

HOLT
BIOSOURCES
TEACHING TRANSPARENCIES

GIVES YOU THE POWER TO ....

- *Simplify complex biological processes*
  Brilliant four-color images lead students step-by-step through processes making even the most challenging topics easier to teach.
- *Bring key structures and relationships to life*
  Clear and highly effective graphics add visual impact to your teaching.

INCLUDES ....

- **238 four-color Teaching Transparencies**
- **71 Transparency Masters** which can be used to make additional transparencies or handouts.

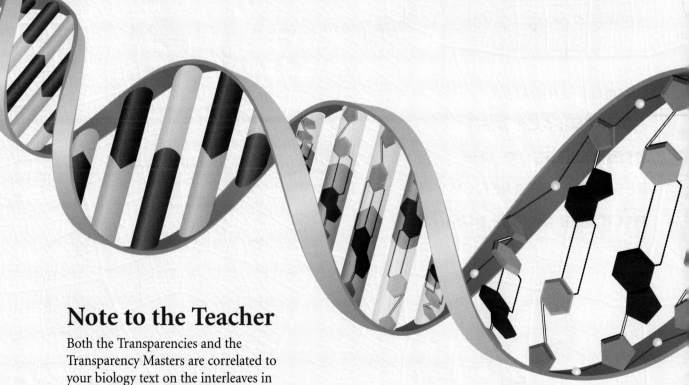

Note to the Teacher

Both the Transparencies and the Transparency Masters are correlated to your biology text on the interleaves in the Annotated Teacher's Edition.

Holt BioSources
Transparency Directory with Teacher's Notes

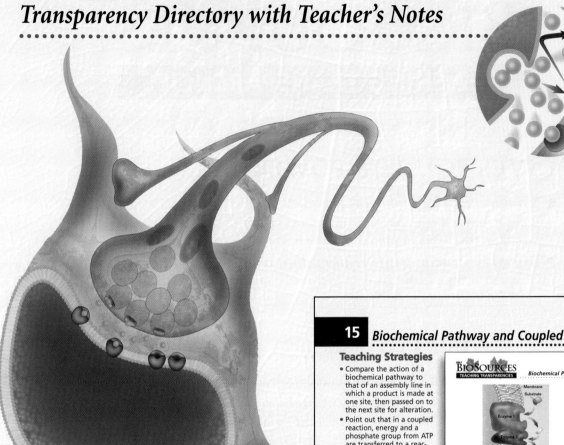

**BRINGS GREATER
CONVENIENCE AND
USEFULNESS TO THE
BEST TRANSPARENCY
PACKAGE IN THE BUSINESS!**

15 *Biochemical Pathway and Coupled Reaction*

Teaching Strategies
- Compare the action of a biochemical pathway to that of an assembly line in which a product is made at one site, then passed on to the next site for alteration.
- Point out that in a coupled reaction, energy and a phosphate group from ATP are transferred to a reactant molecule in an adjacent active site.

What is the role of energy that is transferred in a coupled reaction?
(It activates the reactant.)

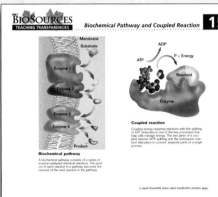

16 *Overview of Cellular Respiration*

Teaching Strategies
- Use this transparency to provide a visual overview of cellular respiration. Emphasize that, as in photosynthesis, each stage of cellular respiration consists of many individual steps. Review the processes of cellular respiration and fermentation. Have students identify the reactants and products of each stage.
- Ask students where in the cell glucose is converted to pyruvate. *(Glucose is converted to pyruvate in cytosol.)*

How much additional ATP results from fermentation?
(none)

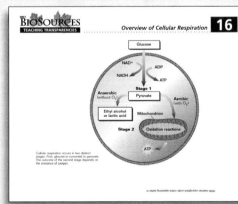

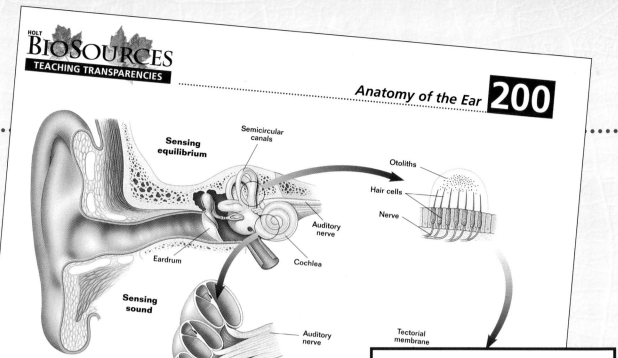

Sensing
equilibrium

Semicircular
canals

Otoliths

Hair cells

Nerve

Auditory
nerve

Eardrum

Cochlea

Sensing
sound

Auditory
nerve

Tectorial
membrane

Electron Transport Chain in Mitochondria 17

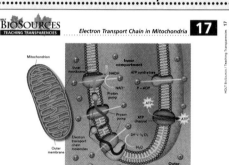

BIOSOURCES
TEACHING TRANSPARENCIES

Electron Transport Chain in Mitochondria 17

Teaching Strategies
- Have students follow an electron's progress (the yellow and red arrows) through the electron transport chain.
- Point out that electron-transport-chain molecules pump protons out of a mitochondrion's inner chamber.
- Ask students what happens when the concentration of protons builds up in the outer compartment. (*Protons are forced back into the inner chamber through ATP synthetase, resulting in ATP production.*)

How does ATP leave the chamber?
(*It leaves through a separate membrane channel.*)

BIOSOURCES
TEACHING TRANSPARENCIES

Photosystems and
Electron Transport 18

Photosystems and Electron Transport 18

Teaching Strategies
- Have students follow an electron's path (the yellow and orange arrows) from the photosystem II reaction center. Point out that one electron-transport-chain molecule pumps protons into the thylakoid space.
- After discussing photosystem I, return to this diagram and emphasize that electrons boosted in photosystem II are passed to photosystem I, not vice versa.

Ask students how ATP is produced in chloroplasts.
(*Protons that build up inside a thylakoid are forced through ATP synthetase.*)

Holt BioSources Transparency Directory

1. The fastest way to preview.

It's easy to see the quality and instructional value of each color transparency and transparency master. The *Transparency Directory* provides reduced, full-color reproductions of every transparency, *plus* teaching suggestions and discussion questions to maximize classroom effectiveness.

2. The best way to plan.

With the *Transparency Directory*, you can conveniently locate the transparencies you'll need to prepare a lesson (and you won't have to transport all the transparencies while you do it). What's more, the directory is organized to correspond to the lesson-by-lesson listings in the Annotated Teacher's Edition Planning Guide.

3. Eye-Popping Quality!

There are two sample transparencies in the back of this directory for you to use with your class—just an example of the quality you get from Holt, Rinehart and Winston.

Contents

HOLT
BIOSOURCES
TEACHING TRANSPARENCIES
TRANSPARENCIES AND
TRANSPARENCY MASTERS

CELL STRUCTURE AND FUNCTION

Transparencies

1 Revision of a Theory: Continental Drift and Plate Tectonics. 1

2 Ionic Bonds. 2

3 Hydrogen Bonds in Water. 3

4 Structure of an Atom. 4

5 Diffusion. 5

6 Osmosis. 6

7 Eukaryotic Cell Structure: Chloroplast, Mitochondrion, and Cytoskeleton 7

8 Eukaryotic Cell. 8

9 Structure of Lipid Bilayer 9

10 Active and Passive Transport. 10

11 Active Transport: Sodium-Potassium Pump. . 11

12 Overview of Photosynthesis 12

13 Energy Flow Through an Ecosystem 13

14 Chromosome and DNA Structure. 13

15 Biochemical Pathway and Coupled Reaction 14

16 Overview of Cellular Respiration 14

17 Electron Transport Chain in Mitochondria . . 15

18 Photosystems and Electron Transport 15

19 Calvin Cycle 16

20 Photosynthesis–Cellular Respiration Cycle . . . 16

21 Glycolysis 17

22 Krebs Cycle 17

23 Two Pathways of Respiration 18

24 Mitosis. 18

Transparency Masters

1A The pH Scale. 1

2A Biologically Important Molecules 2

3A Biologically Important Molecules, continued 3

4A Molecular Structure of ATP 4

5A Surface-Area-to-Volume Ratio 5

6A Endocytosis and Exocytosis. 6

7A Gated Channels. 7

8A Chemical Reactions and Their Products 8

9A Activation Energy and Chemical Reactions 9

10A Three Types of Solutions 10

11A Converting Light Energy to Chemical Energy. 11

12A Comparing Prokaryotes and Eukaryotes. 12

GENETICS

Transparencies

25 Mendel's Experimental Design 19

26 Mendel's Crosses and Results 20

27 Events of Meiosis I. 21

28 Events of Meiosis II 22

29 Crossing Homozygous Pea Plants 23

30 Crossing Heterozygous Pea Plants 24

31 Monohybrid Crosses 25

32 Dihybrid Crosses 26

33 DNA Replication 27

34 Transcription 28

35 Translation: Forming the First Peptide Bond 28

36 Translation: Completing the Protein 29

37 The Genetic Code 29

38 Tracking Inherited Traits (Family Tree) 30

39 Cleaving DNA 30

40 Four Steps in Genetic Engineering. 31

41 Genetic Engineering and Cotton Plants 31

Transparency Masters

25A Griffith's Transformation Experiment 19

26A Removing Introns From Eukaryotic Genes. 20

GENETICS *CONTINUED*

27A Mechanism of the *LAC* Operon 21

28A Chromosome Mutation (Deletion, Duplication, Inversion, Translocation) 22

29A Binary Fission 23

30A Important Genetic Disorders 24

31A Polymerase Chain Reaction 25

32A Making Genetically Engineered Drugs 26

33A Genetically Engineered Medicines 27

EVOLUTION

Transparencies

42 Pasteur's Experiment 32

43 Lerman's Bubble Model 33

44 Darwin's Voyage 34

45 Rate of Decay of Potassium-40 35

46 Miller-Urey Apparatus 36

47 Spontaneous Assembly of RNA 36

48 Evolution of the Horse 37

49 Whale Evolution 37

50 Forelimbs of Vertebrates 38

51 Comparing Hypotheses on Rates of Evolution 38

52 Process of Natural Selection 39

53 Evolutionary Relationships of Anthropoids . . 39

54 Comparison of Chimpanzee and Human Jaws 40

55 Comparison of Gorilla and Australopithecine Skeletons 40

56 Biological Hierarchy of Classification 41

57 Classification of Modern Humans 41

58 Evolutionary Relationships of Plants 42

59 Cladogram of Seven Vertebrates 42

60 Bird Phylogeny and DNA Sequencing 43

61 Theory of Endosymbiosis 43

62 Two Hypotheses on the Evolution of Hominids 44

Transparency Masters

42A Theories of Life's Origins 32

43A Comparing the Hemoglobin Gene Among Species 33

44A Comparing Vertebrate Embryo Development 34

45A Six Kingdoms of Life 35

ECOLOGY

Transparencies

63 Stabilizing, Directional, and Disruptive Selection 44

64 Map of *Homo sapiens* Migration 45

65 Global Distribution of Seven Biomes 46

66 Ecological Races of Seaside Sparrows 47

67 Grassland Food Web 48

68 Food Chain in an Antarctic Ecosystem 48

69 Food Web in an Antarctic Ecosystem 49

70 Ecological Succession at Glacier Bay 49

71 Water Cycle 50

72 Carbon Cycle 50

73 Nitrogen Cycle 51

74 Carbon Dioxide and Average World Temperatures 51

75 Human Population Growth 52

76 Projected Population Growth 52

77 Four Trophic Levels in an Aquatic Ecosystem 53

78 Warbler Foraging Zones 53

79 Three Habitats in a Marine Environment 54

80 Succession in a Developing Ecosystem 54

81 Effects of Acid Rain 55

82 Making an Ecosystem Model 55

83 Effect of Area on Ecosystem Diversity 56

84 Energy Flow Through an Ecosystem 56

85 Amount of Energy at Four Trophic Levels . . . 57

86 Earthworm Niche 57

87 Effect of Competition on an Organism's Niche 58

Contents

Transparency Masters

63A Three Patterns of Population Dispersion . . . 45
64A Two Types of Population Growth 46
65A Causes of Earth's Climate 47

VIRUSES, BACTERIA, PROTISTS, AND FUNGI

Transparencies

88 Two Kingdoms of Prokaryotes 58
89 Three Bacterial Cell Shapes 59
90 Gram Staining 60
91 Tour of a Bacterium 61
92 Bacterial Diseases and Modes of Transmission 62
93 Structures of Adenovirus and Bacteriophage 62
94 Structures of TMV and Influenza Virus 63
95 Tour of a Virus 63
96 Reproductive Cycle of HIV 64
97 Viral Diseases and Modes of Transmission 64
98 Protist Reproduction 65
99 Structure of *Euglena* 65
100 Tour of a Protist (*Paramecium*) 66
101 Life Cycle of *Plasmodium* 66
102 Diseases Caused by Protists 67
103 Reproduction of *Chlamydomonas* 67
104 Reproduction of *Ulva* 68
105 Six Divisions of Algae 68
106 Classification of Fungi 69
107 Structure of a Mycelium 69
108 Life Cycle of a Zygomycete 70
109 Life Cycle of an Ascomycete 70
110 Life Cycle of a Basidiomycete 71
111 Up Close: Mushroom 71

Transparency Masters

88A Comparing Eubacteria With Archaebacteria 59
89A Members of the Kingdom Protista 60
90A Phyla of Protists 61

PLANTS

Transparencies

112 Meal Supplying a Complete Protein 72
113 Major Crop-Producing Regions of the World 73
114 Early Cultivated Plants 74
115 Characteristics of Dicots and Monocots . . . 75
116 Life Cycle of an Angiosperm 76
117 Life Cycle of a Fern 77
118 Life Cycle of a Moss 78
119 Structure of a Leaf 79
120 Plant Cell Structure 80
121 Structure of a Vascular Plant 81
122 A Closer Look at a Leaf 81
123 Structure of Stems 82
124 Structure of Roots 82
125 Structure of Xylem 83
126 Structure of Phloem 83
127 Pressure-Flow Model of Translocation 84
128 Events in Germination 84
129 Secondary Growth 85
130 Photoperiodism in Plants 85
131 Floral Structure 86
132 Cross Section of an Angiosperm Ovary 86
133 Life Cycle of a Conifer 87

Transparency Masters

112A Medicines Originally Derived From Plants 72
113A Phyla of Living Plants 73
114A Phyla of Living Plants, continued 74
115A Life Cycles of Plants 75
116A Modified Leaves 76
117A Modified Stems 77
118A Types of Plants 78
119A Went's Experiment 79
120A Stages of Plant Differentiation 80

ANIMALS

Transparencies

134 Possible Evolutionary Relationships of Major Animal Groups 87

135 The Animal Body: An Evolutionary Journey 88

136 Major Orders of Insects 89

137 Life Cycle of a Monarch Butterfly 90

138 Digestive Tract of a Bee 91

139 Tracheal System of a Beetle 92

140 Insect Diversity 93

141 Anatomy of a Clam 94

142 Open Circulatory System in a Bivalve. 95

143 Three Types of Body Construction. 96

144 Structure of a Sponge. 97

145 Sexual Reproduction of Sponges. 98

146 Radial and Bilateral Symmetries 99

147 Sexual Reproduction of *Obelia* 99

148 Sexual Reproduction of *Aurelia* 100

149 Life Cycle of *Schistosoma* 100

150 Life Cycle of Beef Tapeworm. 101

151 Nephridium Function in Chiton 101

152 Closed Circulatory System in an Earthworm 102

153 Exploration of a Cnidarian 102

154 Reproduction of Cnidarians 103

155 Development of a Cnidarian Embryo 103

156 Development of a Flatworm Embryo. 104

157 Exploration of a Flatworm 104

158 Exploration of a Roundworm. 105

159 Exploration of a Mollusk. 105

160 Exploration of an Annelid. 106

161 Exploration of an Arthropod 106

162 Water Vascular System of the Sea Star 107

163 Exploration of a Lancelet. 107

164 Evolution of Jaws 108

165 Internal Anatomy of a Bony Fish 108

166 Structure of a Gill. 109

167 Evolution of Limbs. 109

168 Major Groups of Fishes 110

169 Life Cycle of a Frog. 110

170 Orders of Amphibians 111

171 Comparison of Heart Structure 111

172 Fish and Amphibian Circulation 112

173 Evolutionary Relationships of Reptiles and Descendants. 112

174 Orders of Living Reptiles. 113

175 Major Orders of Birds 113

176 Major Orders of Birds, continued. 114

177 Major Orders of Mammals 115

178 Major Orders of Mammals, continued. . . . 115

179 Orders of Extinct Reptiles 116

180 Amphibian Lung Structure 116

181 Mammalian Lung Structure 116

182 Avian Lung Structure 117

183 Comparison of Vertebrate Excretory Systems 117

Transparency Masters

134A Phylogenetic Tree of Arthropods 88

135A Incomplete Versus Complete Metamorphosis 89

136A Structure of a Marine Worm 90

137A Evolutionary Tree of Vertebrates 91

138A Swim Bladder in Bony Fish 92

139A Lateral Line in Bony Fish 93

140A Operculum in Fish 94

141A Phylogenetic Tree of Reptiles 95

142A Most Likely Cause of Dinosaur Extinction. 96

143A Structure of a Feather 97

144A Comparison of Coyote and Beaver Skulls 98

HUMAN BIOLOGY

Transparencies

184 Structure of the Human Knee. 118

185 Human Skeleton. 119

186 Structure of Bone Tissue 120

187 Inside the Human Coelom 121

188 Cross Section of Skin 122

189 Three Types of Epithelial Tissue 123

Contents

HUMAN BIOLOGY *CONTINUED*

190 Structure of Skeletal Muscle 124
191 Three Types of Skeletal Tissue 125
192 Contraction of a Muscle 126
193 Three Types of Muscle Tissue 127
194 Structure of a Neuron 128
195 Physiology of a Nerve Impulse 129
196 Peripheral Nervous System 130
197 Action of a Reflex 131
198 Types of Joints 132
199 Structure of the Human Brain 133
200 Anatomy of the Ear 134
201 Structure and Function of
Semicircular Canals 135
202 Anatomy of the Eye 136
203 How the Brain Senses Light 137
204 Endocrine System 138
205 Hypothalamus and Pituitary Gland 138
206 Action of a Steroid Hormone 139
207 Action of a Peptide Hormone 139
208 Maintaining Normal Blood Glucose
Level 140
209 Major Endocrine Glands 140
210 Structure of the Human Heart 141
211 A Closer Look at Blood Vessels 141
212 Systemic Circulation 142
213 Circulation Pathway in the Human
Body 142
214 Circulatory Loops in the Human Body . . . 143
215 Respiratory System in the Human Body . . 143
216 Digestive System in the Human Body 144
217 Overview of Respiration 144
218 Human Kidney Structure 145
219 Cross Section of the Small Intestine 145
220 How a Kidney Machine Works 146
221 Excretory System in the Human Body 146
222 USDA Food Pyramid 147
223 Female Reproductive System 147
224 Male Reproductive System 148
225 Male Hormones and Reproduction 148
226 Ovarian and Menstrual Cycles 149
227 Events Leading to Implantation 149

228 Structure of the Placenta 150
229 How Hormones Work 150
230 Nerves Versus Hormones 151
231 Action of Thyroxine 151
232 Structures of Sperm and Testicle 152
233 Drug Affected Synapse 152
234 Physiology of Addiction 153
235 Inflammatory Response 153
236 Events of the Immune Response 154
237 How a Killer T Cell Recognizes an
Infected Cell 154
238 Known Routes of HIV Transmission 155

Transparency Masters

184A Three Basic Types of Joints 118
185A Electrocardiogram 119
186A Lymphatic System 120
187A Composition of Blood 121
188A Clotting Cascade 122
189A Blood Types 123
190A Nutrition Label 124
191A Vitamins 125
192A Trace Elements 126
193A Digesting Food Molecules 127
194A Major Endocrine Glands and
Hormones 128
195A Major Endocrine Glands and
Hormones, continued 129
196A Events of Human Fetal Development . . . 130
197A Divisions of the Nervous System 131
198A Structure of Representative
Hormones 132
199A How Parathyroid Hormone Affects
Blood Calcium Levels 133
200A Effect of Aldosterone 134
201A Types of Psychoactive Drugs 135
202A Immune System Response Time 136
203A AIDS Cases Among Young Adults 137

 HOLT **BIOSOURCES**
TEACHING TRANSPARENCIES

Revision of a Theory:
Continental Drift and Plate Tectonics

1

Alfred Wegener thought the continents were once one giant continent. Scientists who did not think that the continents could move ridiculed this idea. Many years later, enough evidence was gathered to show his theory partially correct. Continental drift is explained by the theory of plate tectonics.

The continents were once part of a larger continent called Pangaea.

Heat and pressure beneath the Earth's mantle caused the continents to drift apart.

Continental shapes, and fossil and rock evidence indicate that Pangaea existed.

The theory of plate tectonics resolved some of the controversy over the theory of continental drift. As illustrated here, the continents rest on giant plates that slide across the surface of the Earth.

The theory of continental drift was tested many times before it became widely accepted.

HOLT BioSources / Teaching Transparencies 1

HRW material copyrighted under notice appearing earlier in this work.

Teaching Strategies

- Point out Pangaea and review the basic steps of the scientific method.

What evidence supports the theory that Pangaea existed?
(continental shapes, fossils, and rocks)

How did the theory of plate tectonics resolve some of the controversy regarding the theory of continental drift?
(It explained how continents slide across the surface of the Earth.)

- Ask students to name regions of the world where earthquakes are common and to use the diagram to explain why these regions experience this activity. Ask students if they think earthquakes support the theory of continental drift.

HOLT **BIOSOURCES**
TRANSPARENCY MASTER

The pH scale **1A**

pH value	Examples of solutions
0	Concentrated hydrochloric acid
1	
2	Stomach acid / Lemon juice
3	Vinegar, cola, apples
4	Tomatoes
5	Spinach
6	Normal rainwater / Urine
7	Saliva / Blood
8	Sea water
9	Baking soda
10	
11	Household ammonia
12	
13	Oven cleaner, lye
14	Sodium hydroxide (NaOH)

Increasingly acidic / Neutral H+ = OH- / Increasingly basic

HRW material copyrighted under notice appearing earlier in this work.

The pH Scale **1A**

Teaching Strategies

- Emphasize that a low pH signifies high acidity. Conversely, a high pH means low acidity, thus high alkalinity.
- Have students find the pH of most bodily fluids. *(Most bodily fluids have a pH near 7, the neutral point.)* Ask them what would happen if tissues were exposed to extreme pH levels. *(Tissues would be damaged, as when skin comes into contact with hydrochloric acid or sodium hydroxide.)*
- Have students propose why stomach acid typically does not harm the lining of the digestive tract. Have them link this information to the occurrence of peptic ulcers.

2 Ionic Bonds

Teaching Strategies

- Use this transparency to compare ionic bonds with covalent bonds. Ask students how these two types of bonds differ. *(In a covalent bond, the two atoms share electrons. In an ionic bond, one atom gains one or more electrons lost by the other atom.)* Have students discuss why the sodium atom loses an electron to a chlorine atom, instead of the other way around.

- Inform students that substances with ionic bonds often dissolve when placed in water. For example, table salt dissolves in water, producing a solution of sodium ions and chloride ions, or a sodium chloride solution.

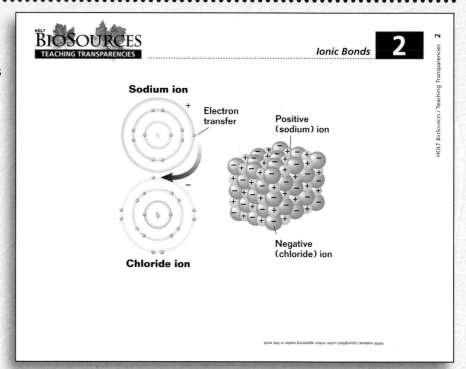

BioSources
TEACHING TRANSPARENCIES

Ionic Bonds 2

Sodium ion

Electron transfer

Positive (sodium) ion

Negative (chloride) ion

Chloride ion

2A Biologically Important Molecules

Teaching Strategies

- Have students identify the three kinds of atoms that make up lipid molecules and carbohydrate molecules *(carbon, hydrogen, and oxygen)*.

How do the proportions of these atoms differ in these two molecules?

(Lipids have more hydrogen atoms bonded to carbon atoms than do carbohydrates, which contain carbon, hydrogen, and oxygen in proportions of 1:2:1.)

- Have students identify the three main portions of a nucleic acid molecule *(the phosphate group, the sugar deoxyribose or ribose, and the nitrogen base)*.

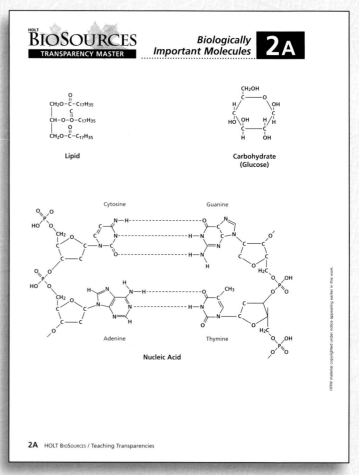

BioSources
TRANSPARENCY MASTER

Biologically Important Molecules 2A

Lipid

Carbohydrate (Glucose)

Cytosine   Guanine

Adenine   Thymine

Nucleic Acid

2A HOLT BioSources / Teaching Transparencies

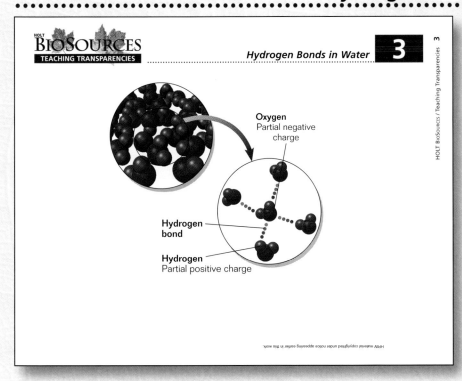

Teaching Strategies

- Have students draw a water molecule on a sheet of paper and label its slightly positive ends *(the two hydrogen atoms)* and its slightly negative center *(the oxygen atom)*. Ask students to describe hydrogen bonding. *(Hydrogen bonding is the weak attraction between the slightly negative region of one molecule and the slightly positive region of another molecule.)*
- Have students draw several more water molecules on the paper and connect them with hydrogen bonds, using the transparency as a model.
- Discuss some of the properties of water that result from hydrogen bonding *(high surface tension, capillarity, and wettable)*.

Biologically Important Molecules, continued 3A

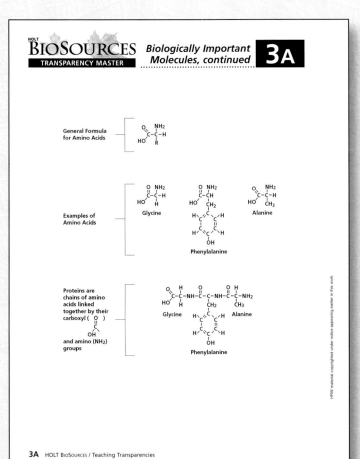

Teaching Strategies

- Have students write the general formula for an amino acid.

What does R represent?
(R represents the side chain, which varies among the 20 possible amino acids.)

- Have students draw the molecular structure of the R groups that make up the amino acids glycine, phenylalanine, and alanine.
- Ask students to describe how amino acids bond to form a polypeptide chain.

What byproduct is released when two amino acids join?
(H₂O)

Teaching Strategies

• Use this transparency when describing atomic structure. Have students identify and describe an atom's nucleus and orbitals. *(The nucleus, the atom's dense core, consists of protons and neutrons. Each proton has a positive charge; neutrons have no charge. The atom's orbitals contain one or more electrons, each with a negative charge.)*

When does an atom have no charge?

(An atom has no charge when the number of protons equals the number of electrons.)

• Display a periodic table of elements. Explain the significance of atomic number. Have students choose an element and illustrate its atomic structure, using the transparency as a model.

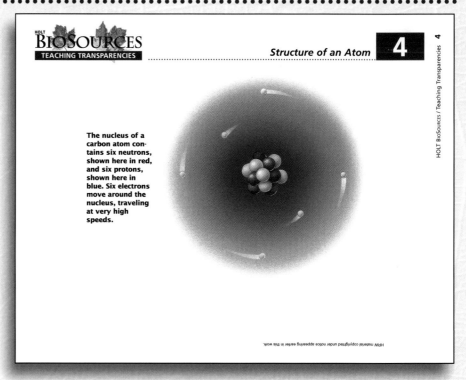

BIOSOURCES
TEACHING TRANSPARENCIES

Structure of an Atom **4**

HOLT BioSources / Teaching Transparencies

The nucleus of a carbon atom contains six neutrons, shown here in red, and six protons, shown here in blue. Six electrons move around the nucleus, traveling at very high speeds.

HRW material copyrighted under notice appearing earlier in this work.

Teaching Strategies

• Ask students to explain why ATP is important in living organisms. *(ATP transports usable energy throughout the cell.)*

• Have student locate the three phosphate groups in ATP. Ask them how energy is released from an ATP molecule *(by removing one of the phosphate groups).*

• Have students draw the structure of the molecule that results when energy is released from an ATP molecule. *(Students should draw the structure of ADP.)*

• Compare the structure of the ATP molecule with the structure of a DNA molecule. Have students point out the features the two molecules share. *(Both molecules contain adenine and a five-carbon sugar.)*

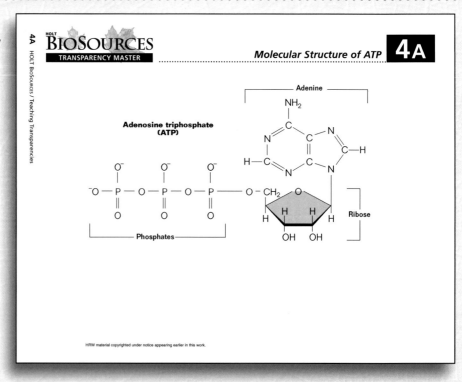

4A HOLT BioSources / Teaching Transparencies

BIOSOURCES
TRANSPARENCY MASTER

Molecular Structure of ATP **4A**

Adenosine triphosphate
(ATP)

Adenine

Phosphates

Ribose

HRW material copyrighted under notice appearing earlier in this work.

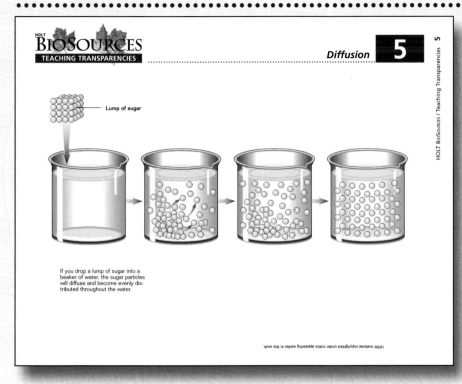

BioSOURCES
TEACHING TRANSPARENCIES   *Diffusion* 5

Lump of sugar

If you drop a lump of sugar into a beaker of water, the sugar particles will diffuse and become evenly distributed throughout the water.

Teaching Strategies

- Review the definition of *diffusion*. Have students hypothesize about the state of matter in which diffusion occurs most quickly. *(Diffusion occurs most quickly in gases, where a large amount of space exists between particles and where there is little chance of collisions that retard movements.)*

- Have students hypothesize about why the rate of diffusion is slower in cold liquids than in warm liquids. *(In these states, molecules are moving faster.)*

- Ask students to describe the conditions under which molecules in living organisms diffuse. *(Molecules are generally in a warm, aqueous solution.)*

Surface-Area-to-Volume Ratio 5A

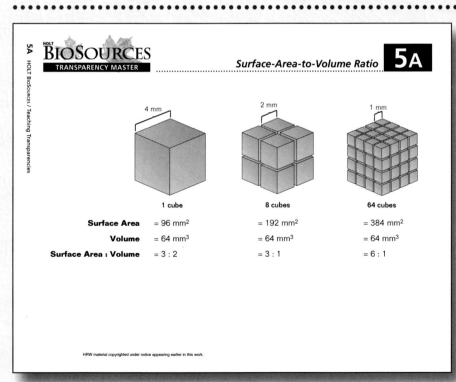

BioSOURCES
TRANSPARENCY MASTER   *Surface-Area-to-Volume Ratio* 5A

	1 cube	8 cubes	64 cubes
Surface Area	= 96 mm²	= 192 mm²	= 384 mm²
Volume	= 64 mm³	= 64 mm³	= 64 mm³
Surface Area : Volume	= 3 : 2	= 3 : 1	= 6 : 1

4 mm   2 mm   1 mm

Teaching Strategies

- Review the formulas for calculating the area of a square and the volume of a cube.

- Lead students through the calculations of volume and surface area for the first cube shown in the transparency. Have students calculate the measurements for the second and third cubes. Be sure they recognize the dramatic increase in surface-area-to-volume ratio as the cube size decreases.

- Discuss the biological importance of surface-area-to-volume ratio.

6 *Osmosis*

Teaching Strategies

- Review the definitions of *osmosis* and *selectively permeable membrane*. (*Osmosis is the diffusion of water across a selectively permeable membrane from a region of high concentration to a region of low concentration.*) Inform students that the unique properties of the solute molecules—their size and electrical charge—cause osmosis to occur.

How can water molecules pass through the membrane illustrated in the transparency?

(The water molecules are small enough to fit through the perforations in the membrane.)

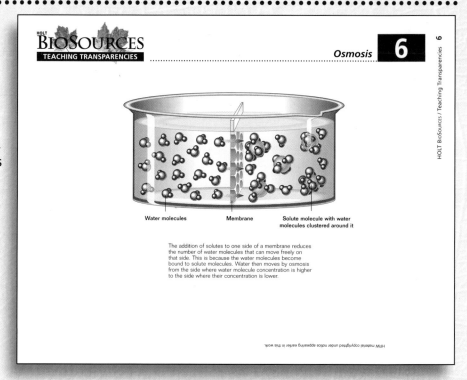

BioSources
TEACHING TRANSPARENCIES

Osmosis 6

Water molecules   Membrane   Solute molecule with water molecules clustered around it

The addition of solutes to one side of a membrane reduces the number of water molecules that can move freely on that side. This is because the water molecules become bound to solute molecules. Water then moves by osmosis from the side where water molecule concentration is higher to the side where their concentration is lower.

HOLT BioSources / Teaching Transparencies 6

6A *Endocytosis and Exocytosis*

Teaching Strategies

- Refer students to the text to find the etymology of *endocytosis* and *exocytosis* and to find definitions for *phagocytosis* and *pinocytosis*. (*Phagocytosis is the cell-engulfing process that results in the formation of large particles or chunks of matter. Pinocytosis is the cell-engulfing process that results in a liquid containing dissolved molecules.*)

- Inform students that in some cases of endocytosis, specific receptors cluster to form a coated pit, which acts to attract certain molecules. This clustering is called receptor-mediated endocytosis and is the process by which cholesterol molecules enter cells.

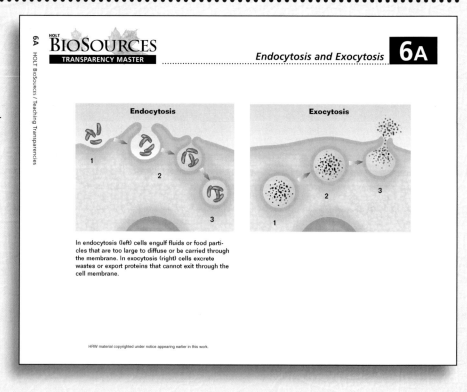

6A HOLT BioSources / Teaching Transparencies

BioSources
TRANSPARENCY MASTER

Endocytosis and Exocytosis 6A

Endocytosis

1
2
3

Exocytosis

1
2
3

In endocytosis (left) cells engulf fluids or food particles that are too large to diffuse or be carried through the membrane. In exocytosis (right) cells excrete wastes or export proteins that cannot exit through the cell membrane.

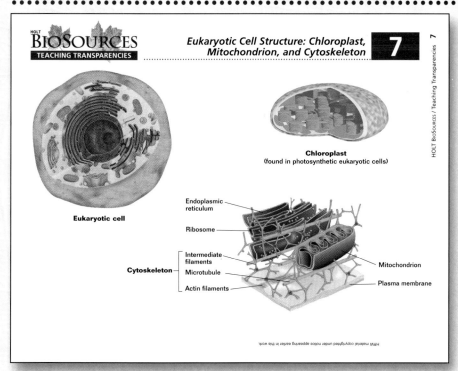

BIOSOURCES
TEACHING TRANSPARENCIES

Eukaryotic Cell Structure: Chloroplast, Mitochondrion, and Cytoskeleton **7**

HOLT BioSources / Teaching Transparencies 7

Chloroplast
(found in photosynthetic eukaryotic cells)

Eukaryotic cell

Endoplasmic reticulum

Ribosome

Intermediate filaments

Cytoskeleton — Microtubule

Actin filaments

Mitochondrion

Plasma membrane

HRW material copyrighted under notice appearing earlier in this work.

Teaching Strategies

- Have students locate the mitochondrion and cytoskeleton inside the eukaryotic cell. Ask them to describe the functions of these two structures.
- Ask students why the chloroplast is not shown inside the eukaryotic cell on the transparency. *(The cell represents an animal cell.)*

In what kinds of eukaryotic organisms are chloroplasts found? *(plants and protists)*

- Be sure students can distinguish a cell's cytoskeleton from its cytoplasm and cytosol. *(Cytoplasm is the mostly fluid environment within the cell membrane; cytosol is the fluid part of the cytoplasm; cytoskeleton is an intracellular complex array of protein fibers.)*

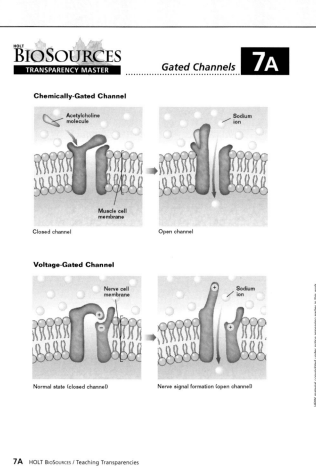

BIOSOURCES
TRANSPARENCY MASTER

Gated Channels **7A**

Chemically-Gated Channel

Acetylcholine molecule

Sodium ion

Muscle cell membrane

Closed channel

Open channel

Voltage-Gated Channel

Nerve cell membrane

Sodium ion

Normal state (closed channel)

Nerve signal formation (open channel)

HRW material copyrighted under notice appearing earlier in this work.

Gated Channels **7A**

Teaching Strategies

- Have students compare the signal that opens a chemically gated channel with the signal that opens a voltage-gated channel. *(The signal for the chemically gated channel is the molecule acetylcholine. The signal for the voltage-gated channel is a change in voltage.)*
- Inform students that many nerve cells that connect to muscle cells release the substance acetylcholine.

What is acetylcholine's function in the body?
(Acetylcholine stimulates ions to flow across a cell membrane. This action changes the membrane's voltage and causes muscle cells to contract.)

Teaching Strategies

- Have students identify the structures that are unique to plant cells *(chloroplast; large central vacuole; cell wall).*

Why do plant cells need both chloroplasts and mitochondria?

(Chloroplasts capture the sun's energy through photosynthesis; mitochondria release this energy to do work in the cell.)

- Have students search in their text for information about other large cytoplasmic organelles found in plant cells, such as chromoplasts and leucoplasts.

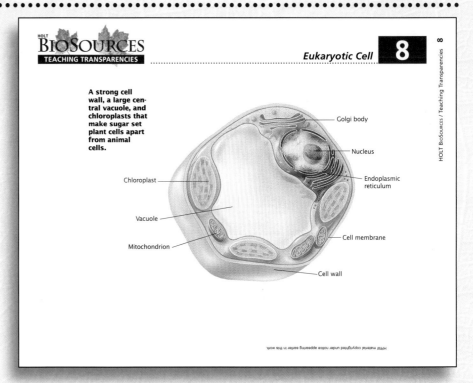

Teaching Strategies

- For each graph, have students compare the amount of energy in the product molecule with the amount of energy in the reactant molecule.
- Point out that the difference between the energy in the reactant and in the product represents the amount of free energy either released or absorbed during the reaction.
- Ask students why the reaction described in the graph on the left is exergonic. *(It releases free energy.)*

Why is the reaction described in the graph on the right endergonic?

(It absorbs free energy.)

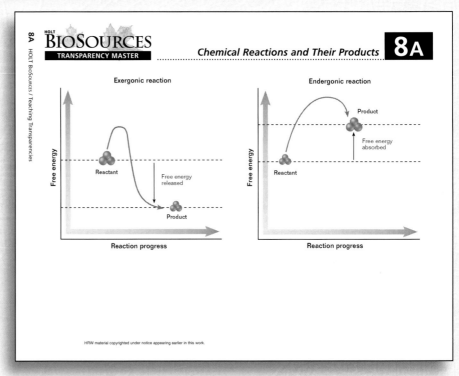

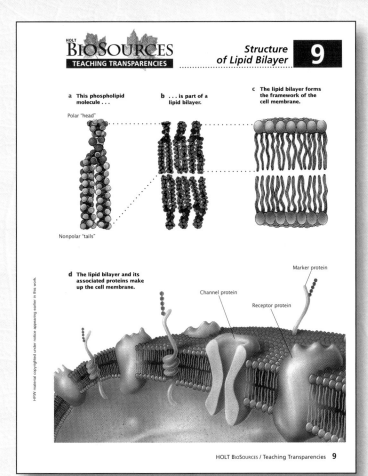

HOLT BIOSOURCES TEACHING TRANSPARENCIES

Structure of Lipid Bilayer **9**

a **This phospholipid molecule . . .**

Polar "head"

Nonpolar "tails"

b **. . . is part of a lipid bilayer.**

c **The lipid bilayer forms the framework of the cell membrane.**

d **The lipid bilayer and its associated proteins make up the cell membrane.**

Marker protein

Channel protein

Receptor protein

HOLT BIOSOURCES / Teaching Transparencies **9**

Teaching Strategies

- Have students describe the structure of a phospholipid molecule. *(A phosphate-containing polar head that is attracted to water is joined to two hydrophobic non-polar tails.)* Ask them how phospholipids are arranged to form a lipid bilayer.
- Ask students to explain how the cell membrane's high permeability to lipids and low permeability to ions is related to its structure.

Activation Energy and Chemical Reactions 9A

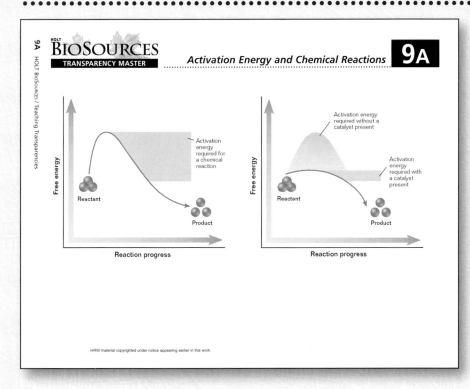

9A HOLT BIOSOURCES / Teaching Transparencies

HOLT BIOSOURCES TRANSPARENCY MASTER

Activation Energy and Chemical Reactions **9A**

Free energy

Reactant

Product

Reaction progress

Activation energy required for a chemical reaction

Free energy

Reactant

Product

Reaction progress

Activation energy required without a catalyst present

Activation energy required with a catalyst present

HRW material copyrighted under notice appearing earlier in this work.

Teaching Strategies

- Tell students that activation energy, which is represented by the hump in each graph line, is the amount of energy that must be added to a system for a reaction to occur.
- Point out the difference in the height of the two humps. Emphasize that a catalyst can greatly reduce the activation energy needed for a reaction but that it cannot eliminate the need for activation energy.
- Ask students if the amount of activation energy needed for a reaction changes the amount of energy given off *(no)*.

Teaching Strategies

• Review the differences between active and passive transport.

What is the energy source for active transport mechanisms? *(ATP)*

• Have students design a graphic organizer that illustrates the transport mechanisms that move materials across a cell membrane. Have them include examples of materials each mechanism transports.

• Have students determine which side of the cell membrane has a higher concentration of solutes (*the outside of the cell*).

How can a cell move material against this gradient? *(It uses active transport.)*

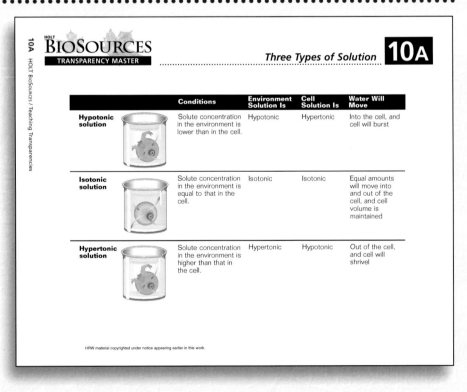

BIOSOURCES
TEACHING TRANSPARENCIES

Active and Passive Transport **10**

HOLT BioSources / Teaching Transparencies 10

10A *Three Types of Solutions*

Teaching Strategies

• Correlate the effects of osmosis on cells subjected to hypotonic, isotonic, and hypertonic solutions with the movement of water. Emphasize the relative solute concentrations that affect the water movement.

• Ask students why it is important for the correct solute concentration to be maintained in the fluid that bathes cells. (*If the solute concentration changes, the solution surrounding the cells can become hypertonic or hypotonic, causing the cells to shrivel or burst.*)

BIOSOURCES
TRANSPARENCY MASTER

Three Types of Solution **10A**

10A HOLT BioSources / Teaching Transparencies

	Conditions	Environment Solution Is	Cell Solution Is	Water Will Move
Hypotonic solution	Solute concentration in the environment is lower than in the cell.	Hypotonic	Hypertonic	Into the cell, and cell will burst
Isotonic solution	Solute concentration in the environment is equal to that in the cell.	Isotonic	Isotonic	Equal amounts will move into and out of the cell, and cell volume is maintained
Hypertonic solution	Solute concentration in the environment is higher than that in the cell.	Hypertonic	Hypotonic	Out of the cell, and cell will shrivel

HOLT
BIOSOURCES
TEACHING TRANSPARENCIES

Active Transport:
Sodium-Potassium Pump **11**

The sodium-potassium pump moves
sodium ions out of cells and potassium
ions into cells.

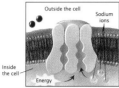

Outside the cell

Sodium
ions

Inside
the cell

Energy

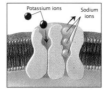

Potassium ions   Sodium
ions

a Sodium ions within the cell fit
precisely into receptor sites on
the channel protein.

b The channel changes shape,
pumping the sodium ions across
the membrane. Potassium ions
outside the cell move into receptor sites.

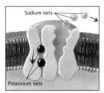

Sodium ions

Potassium ions

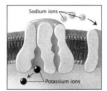

Sodium ions

Potassium ions

c Sodium ions are released and
cannot reenter through this
channel. At the same time, potassium ions are pumped across
the channel into the cell.

d Potassium ions are released
inside the cell. Sodium ions outside the cell, along with sugar
molecules, later enter the cell
through coupled channels.

Teaching Strategies

- Explain to students that the sodium-potassium pump is one of three mechanisms that keep nerve cells functioning. Its role is to restore the balance of ions that is required for nerve-impulse conduction.
- Lead students through the steps of the sodium-potassium pump. Have them create mnemonic devices to help them remember the directions the two ions flow.

Converting Light Energy to Chemical Energy **11A**

11A

HOLT BIOSOURCES / Teaching Transparencies

HOLT
BIOSOURCES
TRANSPARENCY MASTER

Converting Light Energy to Chemical Energy **11A**

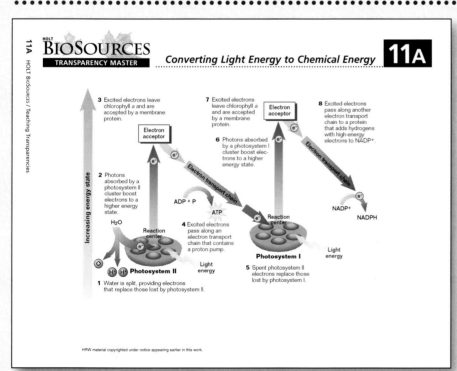

3 Excited electrons leave chlorophyll *a* and are accepted by a membrane protein.

7 Excited electrons leave chlorophyll *a* and are accepted by a membrane protein.

Electron
acceptor

8 Excited electrons pass along another electron transport chain to a protein that adds hydrogens with high-energy electrons to NADP+.

Electron
acceptor

6 Photons absorbed by a photosystem I cluster boost electrons to a higher energy state.

2 Photons absorbed by a photosystem II cluster boost electrons to a higher energy state.

Increasing energy state

ADP + P

ATP

H_2O

Reaction
center

4 Excited electrons pass along an electron transport chain that contains a proton pump.

Reaction
center

NADP+

NADPH

Light
energy

Photosystem I

Light
energy

Photosystem II

1 Water is split, providing electrons that replace those lost by photosystem II.

5 Spent photosystem II electrons replace those lost by photosystem I.

Teaching Strategies

- Use this figure, called a Z diagram, to emphasize the relationship between photosystems I and II.
- Point out that electrons passed to NADPH still contain some of the energy absorbed from light.
- Ask students what replaces the electrons lost by photosystem I. *(The electrons lost by photosystem II replace the electrons lost by photosystem I.)*

What replaces the electrons lost by photosystem II?

(Electrons from the splitting of water replace the electrons lost by photosystem II.)

Teaching Strategies

- Use this transparency to provide a visual overview of photosynthesis. Emphasize that each of the three stages actually consists of many steps.
- Review the process of photosynthesis. Have students identify the reactants and products of each stage. *(The products of Stage 1 are O_2 and high-energy electrons. The products of Stage 2 are hydrogen atoms with high-energy electrons, $NADP^+$, ADP, and P.)*
- Have students write an equation that summarizes the components of the photosynthetic process as shown in this transparency.

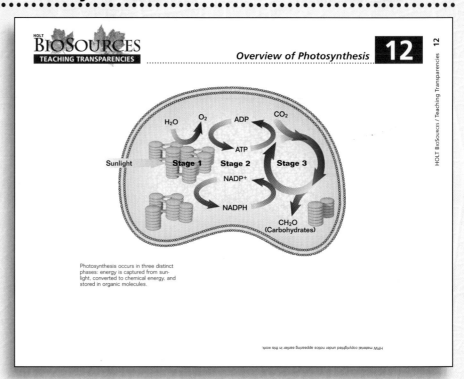

BioSources
TEACHING TRANSPARENCIES

Overview of Photosynthesis **12**

Photosynthesis occurs in three distinct phases: energy is captured from sunlight, converted to chemical energy, and stored in organic molecules.

Teaching Strategies

- Have students compare the magnifications of the photographs of the bacterium and the protist. Which organism is smaller?
- Explain to students that prokaryotes are classified in two kingdoms—Archaebacteria and Eubacteria. Eukaryotes are classified in four kingdoms—Protista, Fungi, Plantae, and Animalia.
- Have students design a graphic organizer that illustrates the similarities and differences between prokaryotes and eukaryotes.

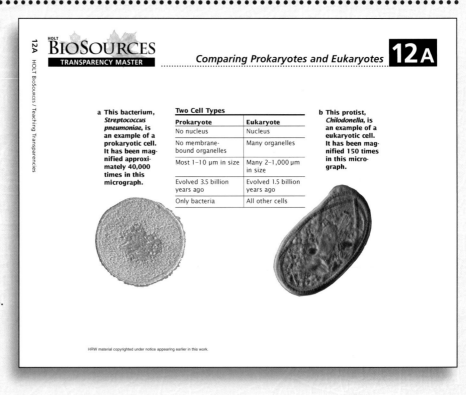

BioSources
TRANSPARENCY MASTER

Comparing Prokaryotes and Eukaryotes **12A**

a This bacterium, *Streptococcus pneumoniae*, is an example of a prokaryotic cell. It has been magnified approximately 40,000 times in this micrograph.

Two Cell Types

Prokaryote	Eukaryote
No nucleus	Nucleus
No membrane-bound organelles	Many organelles
Most 1–10 μm in size	Many 2–1,000 μm in size
Evolved 3.5 billion years ago	Evolved 1.5 billion years ago
Only bacteria	All other cells

b This protist, *Chilodonella*, is an example of a eukaryotic cell. It has been magnified 150 times in this micrograph.

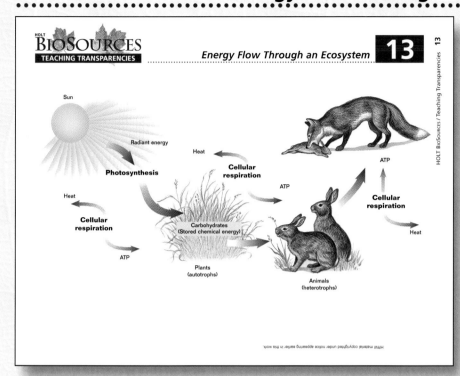

Teaching Strategies

- Follow the path of energy through this simplified food chain. Emphasize that the dissipation of heat causes a loss of energy at each level.

- Stress that through cellular respiration all living systems use the energy in sugars to make ATP within their cells, making the energy available to each cell's machinery.

- Have students identify where the energy in the living system originates *(sunlight)*. Ask them to name the organisms shown in which cellular respiration occurs *(plants, rabbits, and foxes).*

Chromosome and DNA Structure 14

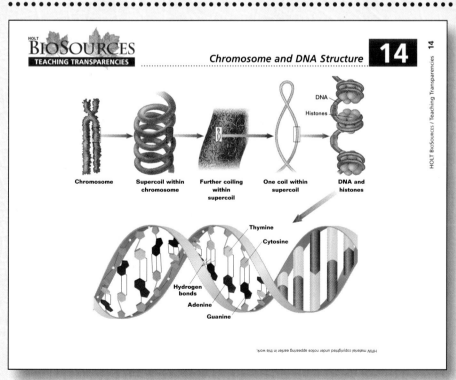

Teaching Strategies

- Use this transparency when describing chromosome structure and formation. Remind students that chromosomes form prior to cell division.

Why is it advantageous for a cell to have its DNA unwound between cell divisions?

(Protein-synthesizing molecules can access the genes in the DNA.)

- Have students hypothesize about what would happen to a cell's hereditary information if chromosomes did not form prior to cell division. *(DNA would become entangled and break, leaving the two new cells without complete copies of genetic information.)*

Teaching Strategies

- Compare the action of a biochemical pathway to that of an assembly line in which a product is made at one site, then passed on to the next site for alteration.

- Point out that in a coupled reaction, energy and a phosphate group from ATP are transferred to a reactant molecule in an adjacent active site.

What is the role of energy that is transferred in a coupled reaction?

(It activates the reactant.)

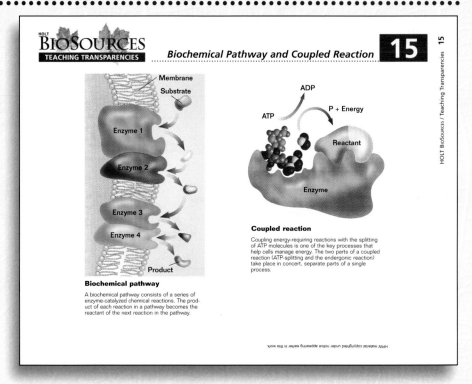

Coupled reaction

Coupling energy-requiring reactions with the splitting of ATP molecules is one of the key processes that help cells manage energy. The two parts of a coupled reaction (ATP-splitting and the endergonic reaction) take place in concert, separate parts of a single process.

Biochemical pathway

A biochemical pathway consists of a series of enzyme-catalyzed chemical reactions. The product of each reaction in a pathway becomes the reactant of the next reaction in the pathway.

Teaching Strategies

- Use this transparency to provide a visual overview of cellular respiration. Emphasize that, as in photosynthesis, each stage of cellular respiration consists of many individual steps. Review the processes of cellular respiration and fermentation. Have students identify the reactants and products of each stage.

- Ask students where in the cell glucose is converted to pyruvate. *(Glucose is converted to pyruvate in cytosol.)*

How much additional ATP results from fermentation?

(none)

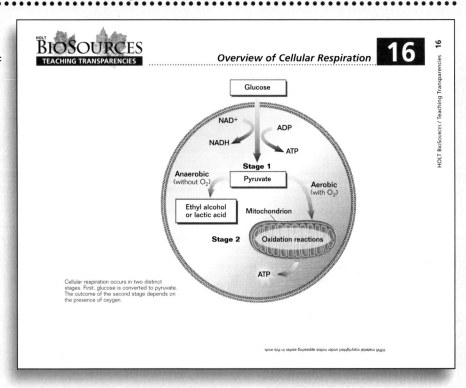

Cellular respiration occurs in two distinct stages. First, glucose is converted to pyruvate. The outcome of the second stage depends on the presence of oxygen.

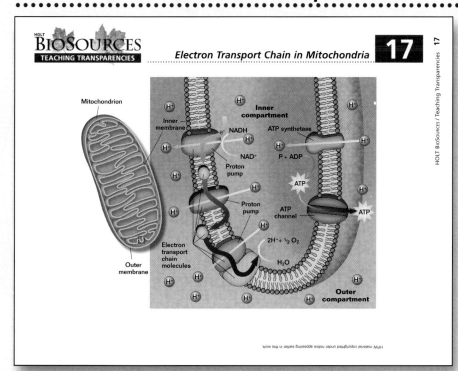

BioSOURCES
TEACHING TRANSPARENCIES
Electron Transport Chain in Mitochondria 17

Teaching Strategies

- Have students follow an electron's progress (the yellow and red arrows) through the electron transport chain.
- Point out that electron-transport-chain molecules pump protons out of a mitochondrion's inner chamber.
- Ask students what happens when the concentration of protons builds up in the outer compartment. *(Protons are forced back into the inner chamber through ATP synthetase, resulting in ATP production.)*

How does ATP leave the chamber?
(It leaves through a separate membrane channel.)

Photosystems and Electron Transport 18

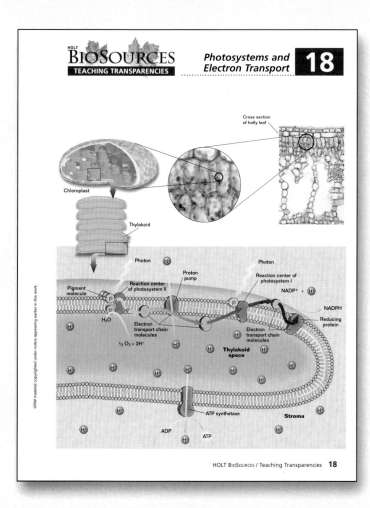

Teaching Strategies

- Have students follow an electron's path (the yellow and orange arrows) from the photosystem II reaction center. Point out that one electron-transport-chain molecule pumps protons into the thylakoid space.
- After discussing photosystem I, return to this diagram and emphasize that electrons boosted in photosystem II are passed to photosystem I, not vice versa.

Ask students how ATP is produced in chloroplasts.
(Protons that build up inside a thylakoid are forced through ATP synthetase.)

Teaching Strategies

- Guide students through this summary of the Calvin cycle. Have them count the total number of carbon atoms present at each step.
- Emphasize that the main product of the Calvin cycle is a three-carbon compound that plants use to make other organic compounds.
- Ask students to describe the function of the Calvin cycle. *(The Calvin cycle incorporates inorganic CO_2 into organic compounds and regenerates the cycle's starting material.)*

How many CO_2 molecules are needed to produce the glucose molecule $C_6H_{12}O_6$?
(6)

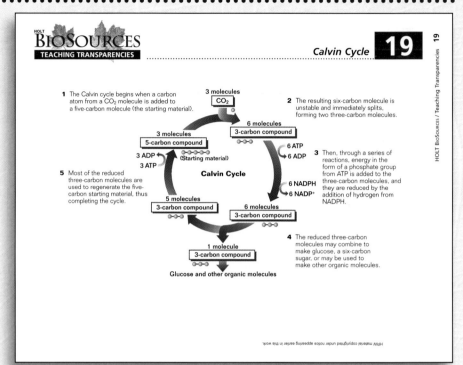

HOLT **BIOSOURCES**
TEACHING TRANSPARENCIES

Calvin Cycle **19**

1 The Calvin cycle begins when a carbon atom from a CO_2 molecule is added to a five-carbon molecule (the starting material).

3 molecules
CO_2

2 The resulting six-carbon molecule is unstable and immediately splits, forming two three-carbon molecules.

6 molecules
3-carbon compound

3 molecules
5-carbon compound

3 ADP
3 ATP
(Starting material)

6 ATP
6 ADP

3 Then, through a series of reactions, energy in the form of a phosphate group from ATP is added to the three-carbon molecules, and they are reduced by the addition of hydrogen from NADPH.

Calvin Cycle

6 NADPH
6 NADP⁺

5 Most of the reduced three-carbon molecules are used to regenerate the five-carbon starting material, thus completing the cycle.

5 molecules
3-carbon compound

6 molecules
3-carbon compound

4 The reduced three-carbon molecules may combine to make glucose, a six-carbon sugar, or may be used to make other organic molecules.

1 molecule
3-carbon compound

Glucose and other organic molecules

Photosynthesis— Cellular Respiration Cycle

20

Teaching Strategies

- Use this figure to visually summarize the two complex metabolic pathways and to show how they are linked. Point out the starting material and products for each process and explain how the products of each process become the starting materials of the other.
- Have students trace the flow of energy from the sun to the chloroplast to the mitochondrion to ATP and heat.
- Ask students if the system illustrated in this figure is open or closed. *(This system is an open system.)*

What scientific law describes the production of heat?
(the second law of thermodynamics)

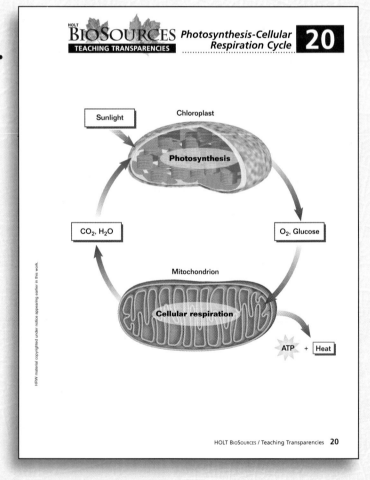

HOLT **BIOSOURCES**
TEACHING TRANSPARENCIES

Photosynthesis-Cellular Respiration Cycle **20**

Sunlight

Chloroplast

Photosynthesis

CO_2, H_2O

O_2, Glucose

Mitochondrion

Cellular respiration

ATP + Heat

BIOSOURCES
TEACHING TRANSPARENCIES

Glycolysis **21**

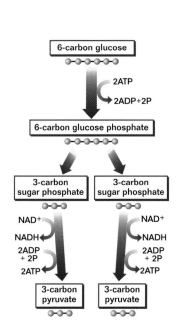

6-carbon glucose

2ATP

2ADP+2P

6-carbon glucose phosphate

3-carbon sugar phosphate	3-carbon sugar phosphate

NAD+ | NAD+

NADH | NADH

2ADP + 2P | 2ADP + 2P

2ATP | 2ATP

| 3-carbon pyruvate | 3-carbon pyruvate |

HFW material copyrighted under notice appearing earlier in this work.

HOLT BioSources / Teaching Transparencies **21**

Glycolysis 21

Teaching Strategies

- Guide students through this summary of glycolysis, emphasizing that glycolysis is an example of a biochemical pathway.
- Point out that cellular respiration uses NAD^+ as an electron carrier instead of the $NADP^+$ used by photosynthesis. Explain that the two molecules are very similar; however, $NADP^+$ has a phosphate group and NAD^+ does not.
- Ask students how many carbon atoms there are at each step of the glycolysis process. *(6)*

How would you calculate the net gain of ATP from glycolysis?
(–2 + 4 = 2)

Krebs Cycle 22

Teaching Strategies

BIOSOURCES
TEACHING TRANSPARENCIES

Krebs Cycle **22**

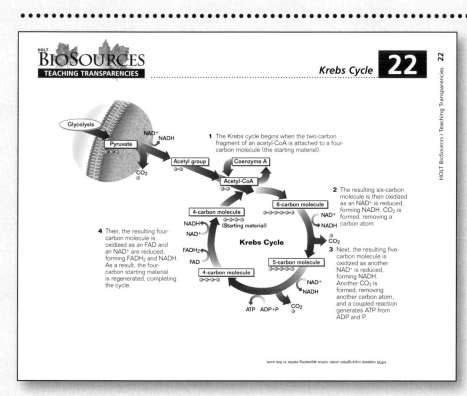

Glycolysis

NAD+ NADH

Pyruvate

CO_2

Acetyl group

Coenzyme A

Acetyl-CoA

1 The Krebs cycle begins when the two-carbon fragment of an acetyl-CoA is attached to a four-carbon molecule (the starting material).

4-carbon molecule

(Starting material)

6-carbon molecule

NAD+

NADH

Krebs Cycle

2 The resulting six-carbon molecule is then oxidized as an NAD+ is reduced, forming NADH. CO_2 is formed, removing a carbon atom.

NADH

NAD+

FADH₂

FAD

4-carbon molecule

5-carbon molecule

CO_2

4 Then, the resulting four-carbon molecule is oxidized as an FAD and an NAD+ are reduced, forming FADH₂ and NADH. As a result, the four-carbon starting material is regenerated, completing the cycle.

ATP ADP+P CO_2

NAD+ NADH

3 Next, the resulting five-carbon molecule is oxidized as another NAD+ is reduced, forming NADH. Another CO_2 is formed, removing another carbon atom, and a coupled reaction generates ATP from ADP and P.

HRW material copyrighted under notice appearing earlier in this work.

- Guide students through this summary of the Krebs cycle. Have students account for the carbon atoms entering and leaving each step. For example, a two-carbon acetyl group is added to a four-carbon starting material to produce a six-carbon molecule.
- Explain that the Krebs cycle must "turn" twice for every glucose molecule that is converted to pyruvate.
- Have students describe the function of the Krebs cycle. *(The Krebs cycle produces electron carriers and ATP while regenerating the cycle's starting material.)*

23 | Two Pathways of Respiration

Teaching Strategies

- Use the flowchart for aerobic pathways to review the steps in the complete oxidation of a glucose molecule. If students notice that the Krebs cycle produces only one ATP, remind them that the Krebs cycle must turn twice for each glucose molecule.
- Use the flowchart for anaerobic pathways to contrast fermentation with the complete oxidation of a glucose molecule.

How many ATPs result from fermentation? *(2)*

How many ATPs result from the complete oxidation of a glucose molecule? *(36)*

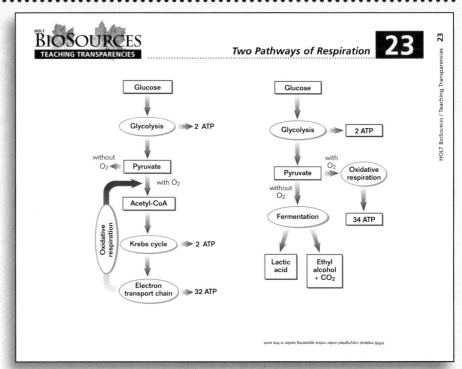

---

24 | Mitosis

Teaching Strategies

- Use this transparency to summarize the events of animal cell reproduction. Remind students that, in reality, these events occur as one continuous process.
- Have students identify the three main stages of a cell's life *(interphase, mitosis, and cytokinesis)*.
- Direct students' attention to the chromosomes in prophase. Ask students how many chromatids are found in a chromosome. *(Each chromosome is made up of two chromatids.)*

During what phase of mitosis do the chromatids separate? *(during anaphase)*

- Ask students to name and describe the three phases of the cell cycle incorporated in interphase *(G₁ phase—cell growth; S phase—DNA replication; G₂ phase—growth and preparation for mitosis)*.

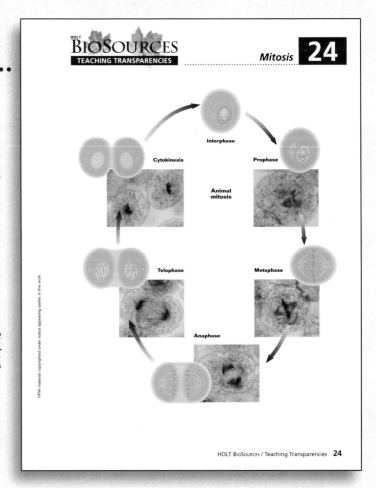

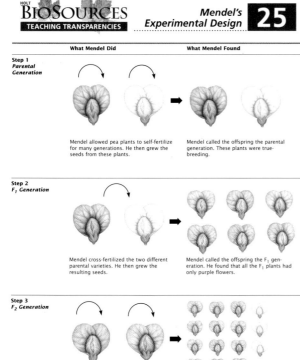

BIOSOURCES
TEACHING TRANSPARENCIES

Mendel's Experimental Design 25

What Mendel Did	What Mendel Found

Step 1
Parental Generation

Mendel allowed pea plants to self-fertilize for many generations. He then grew the seeds from these plants.

Mendel called the offspring the parental generation. These plants were true-breeding.

Step 2
F₁ Generation

Mendel cross-fertilized the two different parental varieties. He then grew the resulting seeds.

Mendel called the offspring the F₁ generation. He found that all the F₁ plants had only purple flowers.

Step 3
F₂ Generation

Mendel allowed the F₁ generation plants to self-fertilize. As before, he grew the seeds from these plants.

Mendel called the offspring the F₂ generation. White flowers reappeared. Of the 929 plants he grew, 705 had purple flowers and 224 had white flowers.

HOLT BioSources / Teaching Transparencies **25**

Mendel's Experimental Design 25

Teaching Strategies

- Use this transparency when introducing Mendel's garden-pea-plant experiments. Review the terms *self-fertilization, cross-fertilization,* and *true-breeding.*

Why did Mendel allow the parental generation pea plants to self-fertilize? *(He wanted to ensure that the parental generation plants were true-breeding.)*

Would the results have been different if one of these plants had not bred true? *(Yes. For example, if the purple-flowering plant had not bred true, white flowers would have appeared in the F1 generation.)*

- Have students calculate the ratio of purple flowers to white flowers in the F2 generation *(3:1).* Tell students that this ratio was significant in Mendel's research.

Griffith's Transformation Experiment 25A

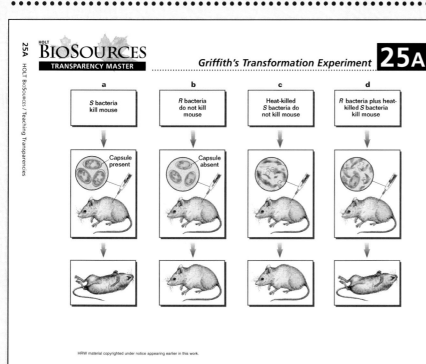

BIOSOURCES
TRANSPARENCY MASTER

Griffith's Transformation Experiment 25A

a	b	c	d
S bacteria kill mouse	*R* bacteria do not kill mouse	Heat-killed *S* bacteria do not kill mouse	*R* bacteria plus heat-killed *S* bacteria kill mouse

Capsule present

Capsule absent

HRW material copyrighted under notice appearing earlier in this work.

Teaching Strategies

- Use this transparency when describing Griffith's transformation experiments. Review the terms *strain, virulent,* and *transformation.*

- Ask students to describe how the S strain of bacteria differs from the R strain of bacteria. *(The S strain bacterium is enclosed within a capsule; the R strain bacterium is not.)*

How might the presence of the capsule render the bacterium harmful to the mouse? *(It may protect the bacterium from attacks by the immune system.)*

26 Mendel's Crosses and Results

Teaching Strategies

- Use this transparency when discussing the results of Mendel's crosses. Review the terms *trait, recessive,* and *dominant.*
- Have students note the two contrasting forms of each trait crossed.

Which form of each trait appears in the F₁ generation?
(dominant form)

- Have students select one pair of traits and design a graphic organizer that illustrates the three steps of Mendel's experiments.

Trait	Dominant vs. Recessive		F₂ Generation Results		Ratio
			Dominant Form	Recessive Form	
Flower color	Purple × White		705	224	3.15:1
Seed color	Yellow × Green		6,022	2,001	3.01:1
Seed shape	Round × Wrinkled		5,474	1,850	2.96:1
Pod color	Green × Yellow		428	152	2.82:1
Pod shape	Round × Constricted		882	299	2.95:1
Flower position	Axial × Top		651	207	3.14:1
Plant height	Tall × Dwarf		787	277	2.84:1

26A Removing Introns from Eukaryotic Genes

Teaching Strategies

- Use this transparency when discussing the process by which introns are removed from exons. Review the terms *intron, exon,* and *gene.* Advise students to think of exons as the segments that get expressed and introns as the segments that intervene.
- Scientific research has shown that introns do not occur in prokaryotic cells. Ask students why this may be advantageous to bacteria. *(Without introns, the DNA can be replicated faster, which would benefit a small, unicellular organism that needs to reproduce rapidly.)*
- Have students design a graphic organizer that compares the events of gene expression in prokaryotes with those in eukaryotes.

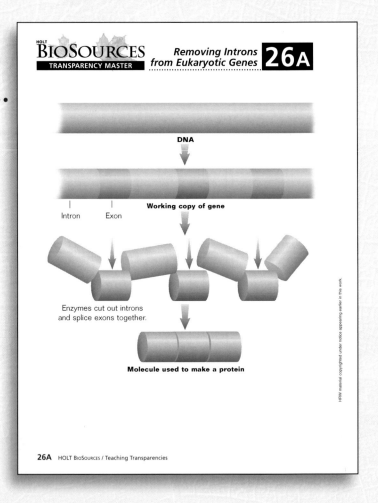

DNA

Working copy of gene

| Intron | Exon

Enzymes cut out introns and splice exons together.

Molecule used to make a protein

26A HOLT BIOSOURCES / Teaching Transparencies

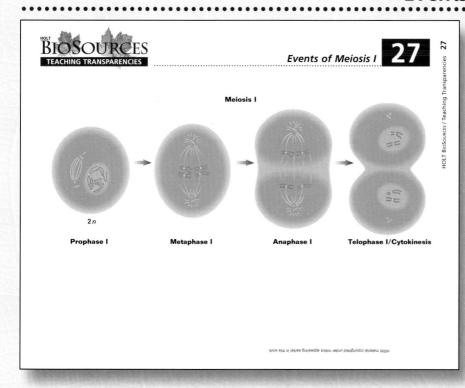

Teaching Strategies

- Use this transparency to review the steps of meiosis I. Review the terms *diploid, chromosome, chromatid, centrioles,* and *spindle fibers.*
- Ask students to describe the outcome of this process. *(Meiosis I reduces the number of chromosomes, resulting in two cells with half the number of chromosomes as the original.)*

How is meiosis different from mitosis?

(Homologous chromosomes pair up in meiosis during prophase before crossing over.)

What is its evolutionary significance of crossing over?

(Crossing over increases the number of different genetic combinations, promoting diversity.)

Mechanism of the LAC Operon 27A

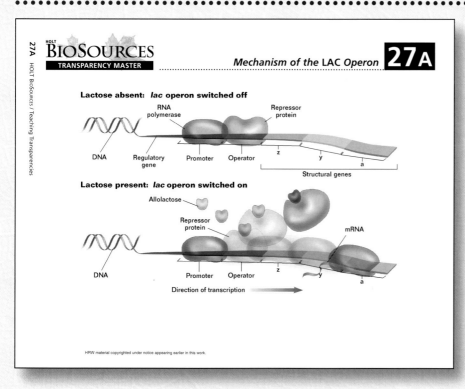

Teaching Strategies

- Use this transparency when discussing the mechanism of the LAC operon. Review the terms *RNA polymerase, repressor protein, promoter,* and *operator.*
- Guide students through the events depicted. Be sure students can distinguish between the two parts of the figure. *(The top half depicts the operon in the absence of lactose; the bottom half illustrates the operon in the presence of lactose.)*
- Ask students to describe how the presence of lactose affects the repressor protein attached to the operator. *(Lactose binds to the repressor protein, which changes its shape. The protein detaches from the operator, and transcription begins.)*

Teaching Strategies

- Use this transparency when describing the events of meiosis II,emphasizing that this process is a continuation of meiosis I. Review the term *haploid*.

- Have students describe the outcome of the process. *(Meiosis II separates chromatids, resulting in four haploid cells.)* Inform students that meiosis usually produces gametes in animals and spores in plants.

- Have students compare the events of meiosis II with mitosis. *(From the standpoint of mechanics, meiosis II is essentially mitotic. However, since the nucleus is haploid, there are no homologous chromosomes to synapse.)*

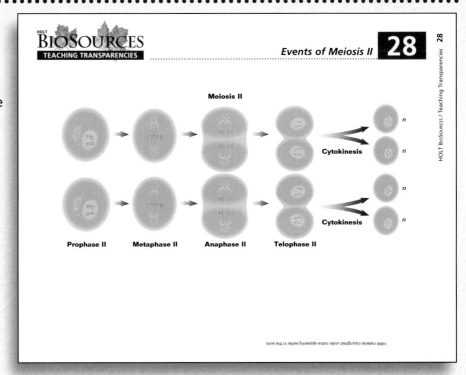

Teaching Strategies

- Use this transparency to illustrate the effects that mutations can have on genes. Review the terms *mutation, deletion, duplication, inversion,* and *translocation*. Explain that the shaded genes in the top row of chromosomes are affected by mutations and that the results are shown in the bottom row.

- Have students describe the difference between duplication and translocation. *(In duplication there are two copies of the duplicated genes. In translocation there is only one copy of each gene.)*

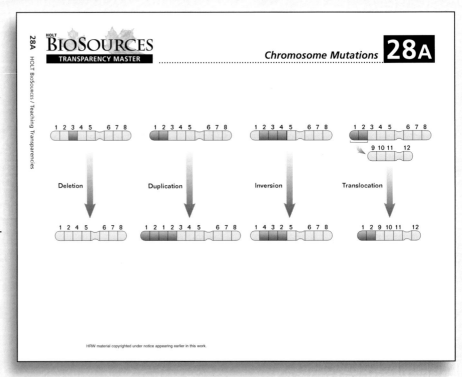

Transparency 29

BIOSOURCES
TEACHING TRANSPARENCIES

Crossing Homozygous Pea Plants 29

This Punnett square illustrates a cross between two true-breeding varieties of garden pea plants. This was Step 2 of Mendel's experiment.

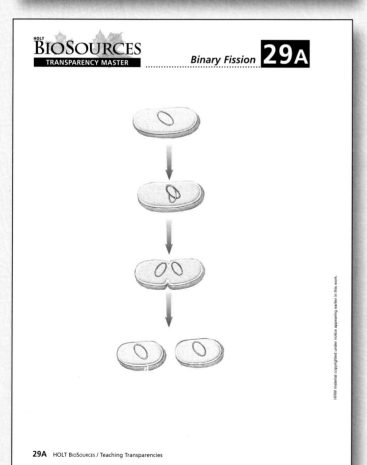

WW

W   Male gametes   *W*

4 The sperm contained within pollen grains produced by this flower . . .

5 . . . can only have a *W* allele because the plant is *WW*.

1 Each gamete, whether male or female, contains a single allele for flower color.

2 The eggs produced by this flower . . .

w

Female gametes

w

ww

3 . . . can only have a *w* allele because the plant is *ww*.

Ww   *Ww*

Ww   *Ww*

6 This is the genotype: *Ww*. All F₁ flowers are heterozygous: *Ww*.

7 This is the phenotype: purple. All F₁ flowers are purple.

HOLT BIOSOURCES / Teaching Transparencies **29**

Crossing Homozygous Pea Plants 29

Teaching Strategies

- Use this transparency when reviewing the details of the second step of Mendel's experiments. Review the terms *homozygous, heterozygous, allele, genotype,* and *phenotype.*
- Review the seven parts of the diagram. Have students explain why a cross between a purple-flowering plant and a white-flowering plant produced only purple-flowering offspring.
- Have students describe the process that forms the haploid gametes produced by the flowers on the sides of the Punnett square. *(Students should describe meiosis.)*

What process gives rise to the diploid genotypes of the flowers inside the boxes of the Punnett square? *(fertilization.)*

- Have students describe the process of fertilization.

BIOSOURCES
TRANSPARENCY MASTER

Binary Fission 29A

29A   HOLT BIOSOURCES / Teaching Transparencies

Binary Fission 29A

Teaching Strategies

- Use this transparency when discussing the events of binary fission. Review the term *asexual reproduction.*
- Have students compare a chromosome in a bacterium with that in a eukaryotic cell. *(A bacterium has one circular chromosome in the cytoplasm, while a eukaryotic cell has many individual chromosomes within a nucleus.)*
- Ask students why bacteria do not undergo mitosis. *(Mitosis is division of the nucleus, which is not present in bacteria.)*

Why is binary fission considered a type of asexual reproduction?

(A single cell is split into two new cells that contain identical genetic information.)

30 Crossing Heterozygous Pea Plants

Teaching Strategies

- Use this transparency when reviewing the details of the third step of Mendel's experiment. Review the terms F_1, F_2, and *heterozygous*.

- Have students express the phenotypes and genotypes of the offspring in this cross as ratios and as percents (*1 homozygous purple: 2 heterozygous purple: 1 homozygous white; 75 percent purple, 25 percent white*).

- Have students review the law of segregation and explain how this Punnett square illustrates Mendel's first law. (*The plants that are crossed each have two hereditary factors for the purple-flower trait. The two factors segregate, or separate, during meiosis to form gametes. When the gametes unite during fertilization, the pairs of factors are restored in the offspring.*)

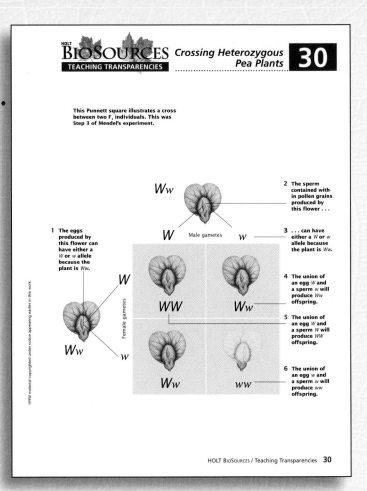

This Punnett square illustrates a cross between two F_1 individuals. This was Step 3 of Mendel's experiment.

1 The eggs produced by this flower can have either a *W* or *w* allele because the plant is *Ww*.

2 The sperm contained within pollen grains produced by this flower . . .

3 . . . can have either a *W* or *w* allele because the plant is *Ww*.

4 The union of an egg *W* and a sperm *w* will produce *Ww* offspring.

5 The union of an egg *W* and a sperm *W* will produce *WW* offspring.

6 The union of an egg *w* and a sperm *w* will produce *ww* offspring.

Male gametes / Female gametes

Ww / *W* / *w* / *WW* / *Ww* / *Ww* / *ww*

HRW material copyrighted under notice appearing earlier in this work.

HOLT BIOSOURCES / Teaching Transparencies **30**

30A Important Genetic Disorders

Teaching Strategies

- Use this transparency to summarize important information about some common genetic disorders. Review the terms *symptom, defect, dominant,* and *recessive*.

- Inform students that phenylketonuria, Tay-Sachs disease, cystic fibrosis, and sickle cell anemia are a few of the hundreds of single-gene disorders. These disorders result from mutations in individual genes that prevent recessive alleles from producing the normal enzyme or protein.

- Ask students what this information means for individuals homozygous for the recessive allele. (*These individuals cannot produce the normal enzyme or protein and will develop the disorder's symptoms.*)

30A HOLT BIOSOURCES / Teaching Transparencies

Disorder	Symptom	Defect	Dominant or Recessive	Frequency Among Human Births
Cystic fibrosis	Mucus clogs lungs, liver, and pancreas; usually don't survive to adulthood	Failure of chloride ion transport mechanism	Recessive	1/1,800 (whites)
Sickle cell anemia	Poor blood circulation	Abnormal hemoglobin molecules	Recessive	1/1,600 (African-Americans)
Tay-Sachs disease	Deterioration of central nervous system in infancy; usually don't survive to adulthood	Defective form of enzyme hexosaminidase A	Recessive	1/1,600 (Jews)
Phenylketonuria	Failure of brain to develop in infancy; usually don't survive to adulthood (if untreated)	Defective form of enzyme phenylalanine hydroxylase	Recessive	1/18,000
Hemophilia	Failure of blood to clot	Defective form of blood clotting factor IX	Sex-linked recessive	1/7,000
Huntington's disease	Gradual deterioration of brain tissue in middle age; shortened life expectancy	Production of an inhibitor of brain cell metabolism	Dominant	1/10,000
Muscular dystrophy	Wasting away of muscles; shortened life expectancy	Muscle fibers degenerate and atrophy	Sex-linked recessive	1/10,000

HRW material copyrighted under notice appearing earlier in this work.

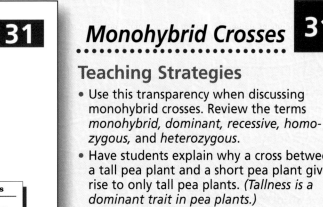

Teaching Strategies

- Use this transparency when discussing monohybrid crosses. Review the terms *monohybrid, dominant, recessive, homozygous,* and *heterozygous.*
- Have students explain why a cross between a tall pea plant and a short pea plant gives rise to only tall pea plants. *(Tallness is a dominant trait in pea plants.)*
- Have students explain why a cross between two black heterozygous rabbits gives rise to black rabbits as well as brown rabbits. Which step of Mendel's experiment does this cross reflect? *(Step 3)*
- Have students construct a monohybrid Punnett square that shows the results of a cross between two heterozygous round-pea-producing plants.

Polymerase Chain Reaction **31A**

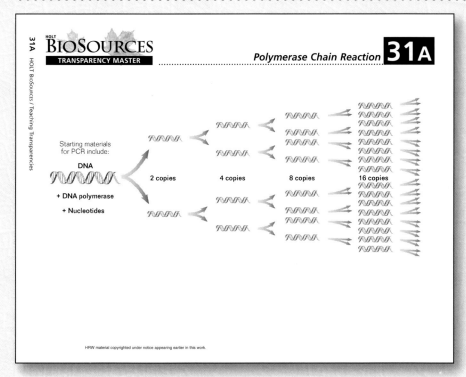

Teaching Strategies

- Use this transparency to illustrate how the polymerase chain reaction technique can be used to produce millions of copies of DNA in just a few hours. Review the terms *DNA polymerase* and *nucleotide.*
- Inform students that the DNA polymerase used in PCR procedures comes from a species of bacterium adapted to life in the near-boiling waters of hot springs. Have students determine why a heat-stable polymerase is necessary for this technique. *(The DNA to be copied is heated to separate the strands, and then each strand is copied.)*
- Have students describe the process DNA undergoes during PCR. *(Students should describe DNA replication.)*

Dihybrid Crosses

Teaching Strategies

- Use this transparency when discussing dihybrid crosses. Review the concepts of *dihybrid* and *independent assortment*.

- Have students explain why a dihybrid cross requires a Punnett square with 16 boxes. *(In a dihybrid cross, pairs of alleles can combine in 16 different ways.)*

- Have students identify the genotypes and phenotypes of the offspring in each Punnett square.

- Have students review the law of independent assortment and explain how these Punnett squares illustrate Mendel's second law. *(Mendel's law of independent assortment states that during gamete formation, pairs of alleles separate independently. This law is illustrated in the four possible gamete combinations listed on each side of the Punnett square.)*

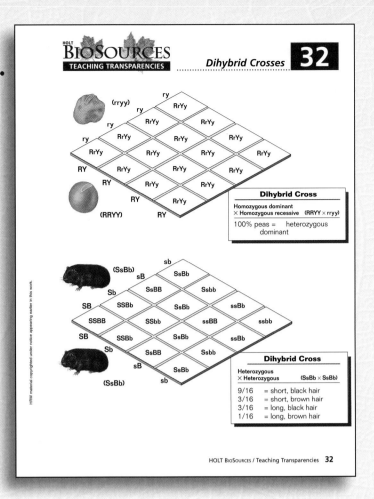

BIOSOURCES
TEACHING TRANSPARENCIES
Dihybrid Crosses 32

Dihybrid Cross

Homozygous dominant
× Homozygous recessive (RRYY × rryy)

100% peas = heterozygous dominant

Dihybrid Cross

Heterozygous
× Heterozygous (SsBb × SsBb)

9/16 = short, black hair
3/16 = short, brown hair
3/16 = long, black hair
1/16 = long, brown hair

HOLT BioSources / Teaching Transparencies **32**

Making Genetically Engineered Drugs

Teaching Strategies

- Use this transparency to illustrate the process used to make genetically engineered pharmaceutical products. Review the term *plasmid*.

- Lead students through the four major steps in this illustration. Have them compare this example with the steps in the Cohen and Boyer experiment.

- In small groups, have students design a genetic engineering experiment that produces human growth hormone to treat a form of dwarfism.

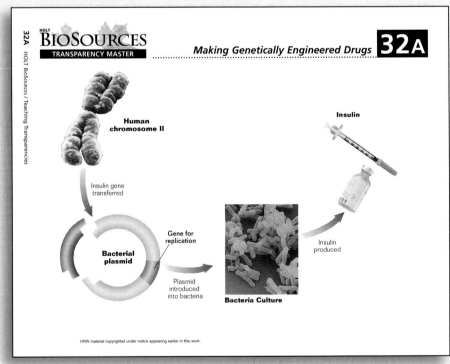

BioSources
TEACHING TRANSPARENCIES
DNA Replication **33**

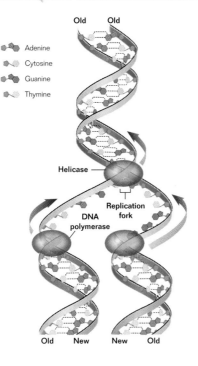

Old   Old

- Adenine
- Cytosine
- Guanine
- Thymine

Helicase

Replication
fork
DNA
polymerase

Old   New   New   Old

HRW material copyrighted under notice appearing earlier in this work.

HOLT BioSources / Teaching Transparencies **33**

DNA Replication 33

Teaching Strategies

- Use this transparency when discussing DNA replication. Review the terms *replication fork*, *DNA polymerase*, and *template*. Be sure that students can identify the original double helix and the two new helices being synthesized.
- Lead students through the steps of replication. Ask them during which phase of the cell cycle DNA replication occurs *(during the S phase)*.

What process typically follows DNA replication?
(mitosis)

- Explain to students that for DNA to be the genetic material, it had to contain information to copy itself. The Watson-Crick model of DNA, revealed in 1953, was successful in part because it immediately suggested a mechanism for replication.

BioSources
TRANSPARENCY MASTER
Genetically
Engineered Medicines **33A**

Product	Examples and Uses
Colony-stimulating factors	Growth factors that stimulate white blood cell production; used to treat immune system deficiencies and to fight infections
Erythropoetin	Stimulates red blood cell production; used to treat anemia in individuals with kidney diseases
Growth factors	Stimulate differentiation and growth of various cell types; used to promote wound healing
Human growth hormone	Used as a treatment for dwarfism
Interferons	Interfere with reproduction of viruses; also used to treat some cancers
Interleukins	Activate and stimulate different classes of white blood cells; can be used in treating wounds, HIV infections, cancer, immune deficiencies

HRW material copyrighted under notice appearing earlier in this work.

Genetically Engineered Medicines 33A

Teaching Strategies

- Use this transparency to exemplify genetically engineered medicines available on the market. Review terms such as *hormone*, *interferons*, and *interleukins*.
- Ask students how interferons work within the immune system. *(The body uses interferons to fight viral infections by helping uninfected cells produce antiviral proteins. They are also used to treat some cancers.)*
- Ask students how interleukins work. *(They stimulate the inflammatory response and T-cell proliferation and they help to produce the symptoms associated with a fever—high body temperature, lack of appetite, aches, pains, and feelings of malaise.)*

What are some uses for interleukins?
(They are used in treating wounds, HIV infection, cancer, and immune deficiencies.)

Teaching Strategies

- Use this transparency when describing the events of transcription. Review the terms *nucleotide, RNA polymerase, template strand*, and *RNA*. Remind students that in RNA, the base uracil is substituted for thymine.

- Have students draw a sequence of 10 nucleotide bases in a DNA strand on a piece of paper, using the transparency as a model. Afterward, have students trade papers. Instruct them to draw the transcribed sequence of their partner's DNA model.

- Have students design a graphic organizer that compares with the events of transcription to those of DNA replication.

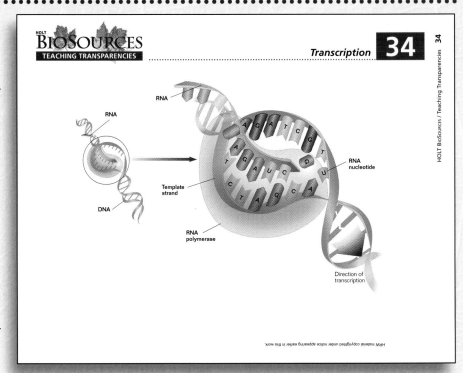

Teaching Strategies

- Use this transparency when describing the structure of tRNA and the structure of a ribosome. Identify these molecules as the tools of translation. Be sure students can identify a ribosome's large and small subunits as well as its A and P sites. Review the terms *codon, anticodon*, and *amino acid*.

- Explain how tRNA and a ribosome align to read the codons along a strand of mRNA. Ask students which site on the large ribosomal subunit gets filled with an amino acid first *(the P site)*.

- Ask students how a ribosome knows where to bind along a strand of mRNA.

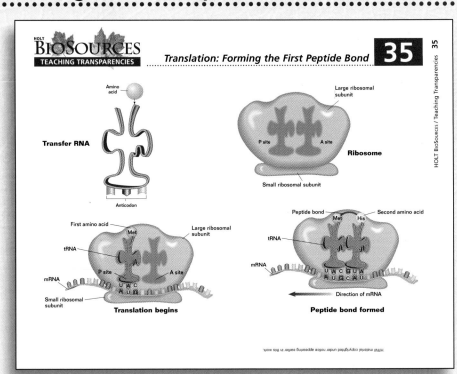

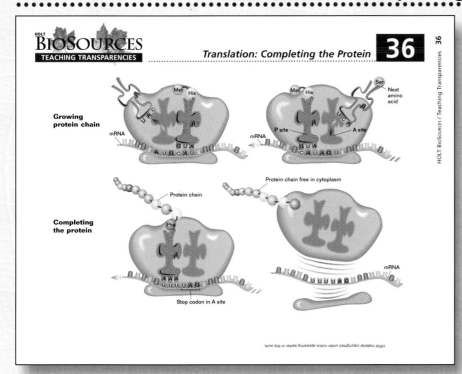

Teaching Strategies

- Use this transparency when describing how a new codon shifts into the A site. Ask students to explain how a strand of mRNA moves across the ribosome during translation. *(The tRNA molecule in the P site detaches, and the tRNA molecule in the A site shifts down to take its place. As a result, a new codon is present in the P site.)*
- Have students identify the stop codon in the mRNA strand in the transparency. Remind them that there are two stop codons—UAA and UAG.

The Genetic Code 37

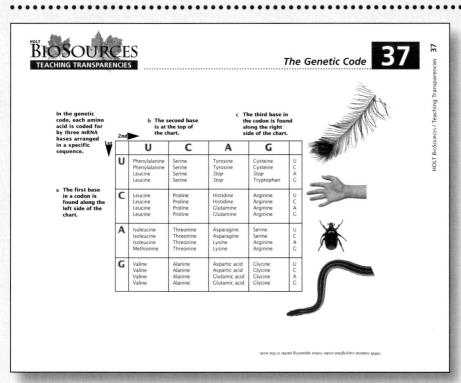

Teaching Strategies

- Read aloud the names of the 20 amino acids in the genetic code key to teach students their correct pronunciations.
- Familiarize students with the procedures for deciphering the genetic code key. Give them a list of amino acids and a list of triplet codons to decipher.
- Ask students why some codons, such as ACU, ACC, ACA, and ACG, can be considered synonyms. *(They all specify the amino acid threonine.)*
- Inform students that the concept of the triplet codon was discovered by Francis Crick and Sydney Brenner in 1961. Their discovery spurred an intensive effort to construct a key that would give meaning to each triplet codon.

38 | Tracking Inherited Traits (Family Tree)

Teaching Strategies

- Use this pedigree when explaining the inheritance of albinism throughout five generations of a family. Emphasize that females are represented by circles and males are represented by squares. Review the terms *sex-linked* and *autosomal*.

- Have students study the pedigree to determine whether albinism is sex-linked or autosomal. Is it dominant or recessive? *(Because albinism appears in both sexes, it is most likely an autosomal trait. And because individuals with albinism have heterozygous parents who are normal, it is a recessive trait.)*

- Have students construct Punnett squares that show the results of each cross in the pedigree.

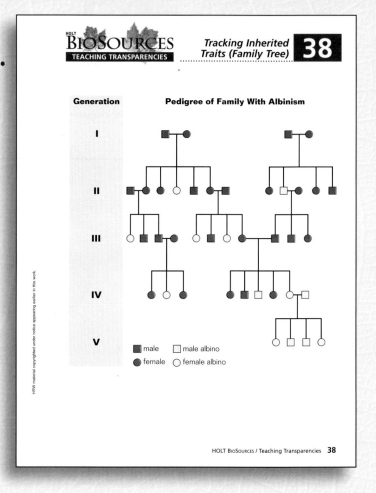

---

39 | Cleaving DNA

Teaching Strategies

- Use this transparency when discussing the techniques scientists use to cleave DNA for a genetic engineering experiment. Review the terms *palindrome*, *EcoRI*, *ligase*, and *recombinant DNA*.

- Have students explain why the single-stranded tails created by restriction enzymes are called sticky ends. *(They can bond with any other complementary single strand of DNA.)*

- Tell students that in DNA there is one chance in four that a given base in a sequence is G, one chance in four that it is an A, and so on. Have them calculate the probability that the sequence GAATTC will occur in this order along a chromosome ($\_ \times \_ \times \_ \times \_ \times \_ \times \_ = 1/4{,}096$).

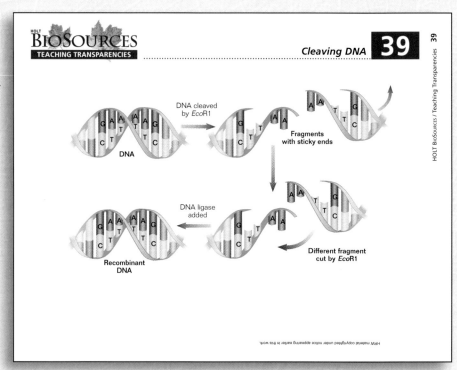

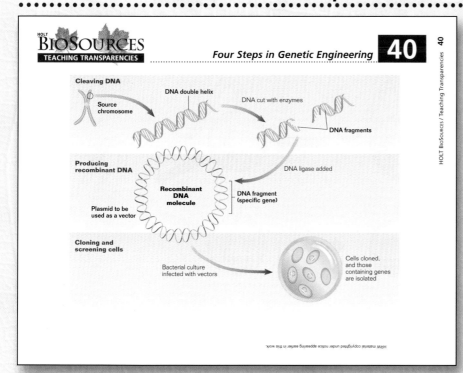

BIOSOURCES
TEACHING TRANSPARENCIES

Four Steps in Genetic Engineering **40**

HOLT BioSources / Teaching Transparencies 40

Cleaving DNA

Source chromosome

DNA double helix

DNA cut with enzymes

DNA fragments

Producing recombinant DNA

DNA ligase added

Recombinant DNA molecule

DNA fragment (specific gene)

Plasmid to be used as a vector

Cloning and screening cells

Bacterial culture infected with vectors

Cells cloned, and those containing genes are isolated

Teaching Strategies

- Use this transparency when outlining the four basic steps of a typical genetic engineering experiment. Review the terms *recombinant DNA, plasmid, vector, cloning*, and *screening*.
- Have students explain why restriction enzymes are considered "essential tools" for genetic engineering experiments. *(They do the work of cutting the DNA at specific sites.)*
- Using the transparency as a guide, have students design a genetic engineering experiment that will enable a dairy cow to produce chocolate milk.

Genetic Engineering and Cotton Plants **41**

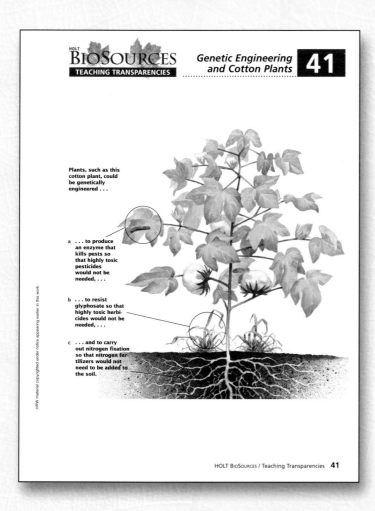

BIOSOURCES
TEACHING TRANSPARENCIES

Genetic Engineering and Cotton Plants **41**

Plants, such as this cotton plant, could be genetically engineered . . .

a . . . to produce an enzyme that kills pests so that highly toxic pesticides would not be needed, . . .

b . . . to resist glyphosate so that highly toxic herbicides would not be needed, . . .

c . . . and to carry out nitrogen fixation so that nitrogen fertilizers would not need to be added to the soil.

HOLT BioSources / Teaching Transparencies **41**

Teaching Strategies

- Use this transparency when discussing the ways in which genetic engineering is improving agricultural products. Review the terms *pesticide, herbicide, glyphosate*, and *nitrogen fixation*.
- Have students explain the techniques used to produce each genetic alteration shown in the illustration.
- In small groups, have students devise an experiment that would enable a cotton plant to produce green-colored cotton fibers.

Teaching Strategies

- Use this transparency when discussing Pasteur's experiments. Review the terms *microorganism, spontaneous generation,* and *biogenesis.*

Why did Pasteur first heat the broth in the flask? *(to kill any microorganisms it might contain)*

Why did he use a flask with a curved neck? *(to trap any microorganisms that might enter the flask)*

- Have students recall mid-nineteenth-century beliefs about life's origins. Ask students how the theory of biogenesis challenged spontaneous generation.

- Have students explain how Pasteur's experiments supported the theory of biogenesis.

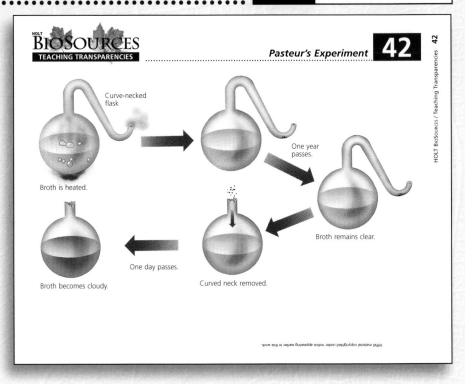

42A Theories of Life's Origins

Teaching Strategies

- Use this transparency when discussing the theories of life's origins. Review the concepts of *organic molecule, inorganic molecule,* and *self-replication.*

- Walk students through the graphic organizer that summarizes the spontaneous origin hypothesis. Point out that sequences of chemical reactions that resulted in increasingly complex organic molecules might have led to life's origin.

- Guide students through the graphic organizer that summarizes the self-replication origin hypothesis. Point out that self-replicating RNA might have arisen before proteins and then catalyzed the formation of the first proteins.

- Ask students what details of life's origins are missing from these diagrams *(origins of DNA, heredity, and cells).*

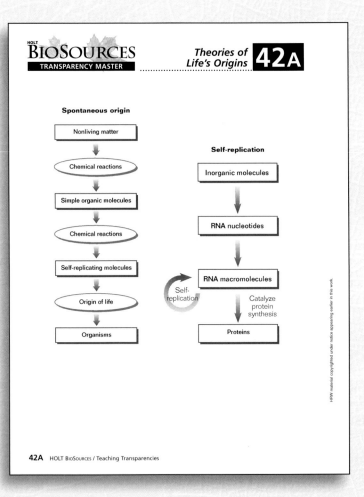

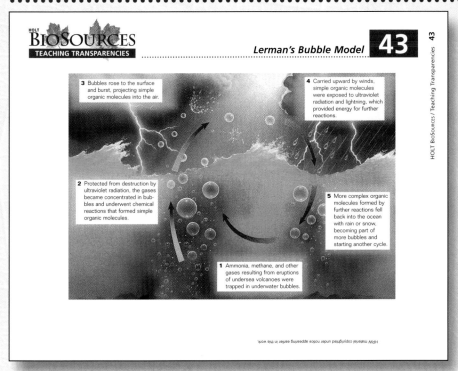

Teaching Strategies

- Walk students through this model, which explains the origin of life's basic chemicals. Point out that the bubbles could act as small reaction vessels by concentrating reactants. Ask students where the energy for these reactions originated *(UV light, lightning, and heat from volcanic eruptions).*

- Be sure students have an understanding of Oparin's primordial soup model. *(Oparin suggested that Earth's oceans were once a vast primordial soup in which life's basic chemicals formed spontaneously in chemical reactions activated by energy from radiation, lightning, and volcanic eruptions.)*

Comparing the Hemoglobin Gene Among Species 43A

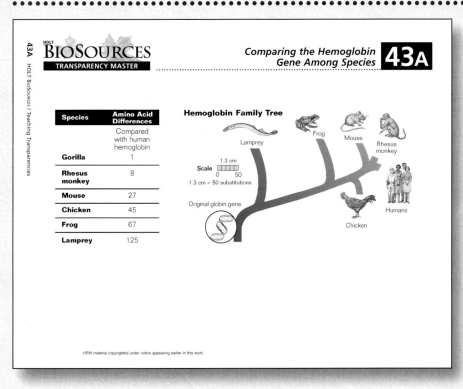

Teaching Strategies

- Use this transparency when discussing how scientists use molecular data to produce family trees. Review the terms *gene, mutation,* and *hemoglobin.*

- Have students compare the table of amino acid differences in species with the hemoglobin family tree. Point out that the table contains data obtained by comparing amino acid sequences, while the family tree contains data obtained by comparing nucleotide sequences.

- Ask students if a family tree produced with the amino acid data in the table would show the same relationships as the family tree based on nucleotide substitutions *(yes).* Why? *(A nucleotide sequence determines an amino acid.)*

Teaching Strategies

- Use this transparency when discussing the route Darwin traveled during his five-year voyage around the world. Be sure students can identify the continents and oceans shown on the map.
- Guide students along the *Beagle's* route.

What was the purpose of this voyage?
(to survey the coast of South America and to visit several Pacific islands)

- In small groups, have students design a travel brochure that describes a sailing trip around the world using the route Darwin traveled.

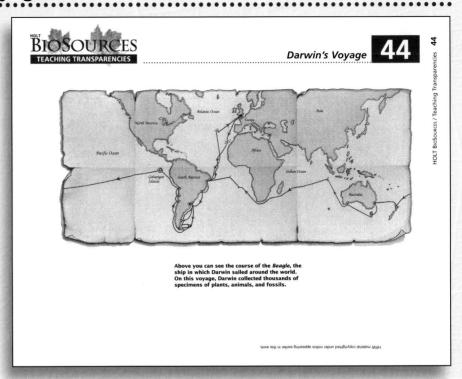

Above you can see the course of the *Beagle,* the ship in which Darwin sailed around the world. On this voyage, Darwin collected thousands of specimens of plants, animals, and fossils.

Teaching Strategies

- Use this transparency when discussing the development of vertebrate embryos. Review the terms *vertebrate, homologous structure,* and *vestigial structure.*
- Ask students to identify the most basic vertebrate characteristic *(the backbone).*

In which organisms do pharyngeal pouches persist the longest? *(in fish)*

Which of these organisms does not have a visible tail as an adult? *(humans)*

- Tell students that different sets of genes are activated as an embryo develops and that the most basic characteristics of an organism appear the earliest in its development.

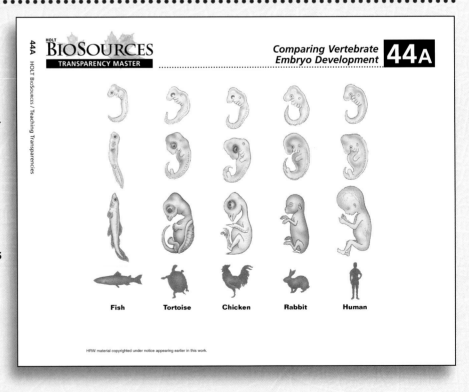

Fish   Tortoise   Chicken   Rabbit   Human

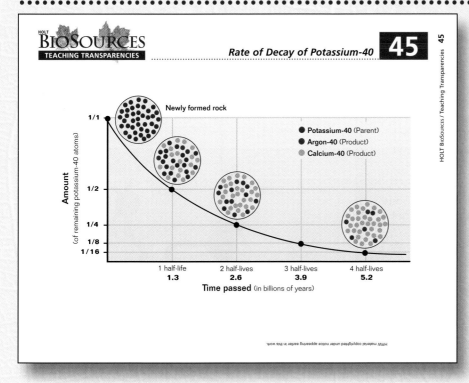

Teaching Strategies

- Use this transparency when discussing the rate of decay in radioisotopes. Review the terms *radiometric dating*, *radioisotopes*, and *half-life*.
- Point out that potassium-40 has two decay products—calcium-40 and argon-40. Emphasize that 50 percent of potassium-40 atoms decay to one or the other of these products during each half-life. Have students use this graph to estimate the age of rocks that contain potassium-40.
- Have students determine the age of a rock that had 1 g of potassium-40 when it formed and has 0.5 g now *(about 1.3 billion years old)*.

If the rock has only 0.125 g of potassium-40, how many half-lives have passed? *(3)*

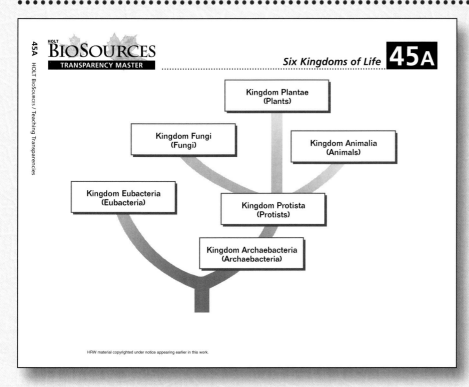

Teaching Strategies

- Use this transparency when discussing the evolutionary relationships of the six kingdoms of life.
- Tell students that every organism on Earth can be placed in one of these six kingdoms. Have them hypothesize why viruses are not included in any of the kingdoms. *(Most biologists do not consider viruses to be living organisms.)*
- Have students name the kingdoms that contain organisms that are prokaryotic *(archaebacteria and eubacteria)*.

What kind of cells make up the organisms contained in the remaining four kingdoms? *(eukaryotic cells)*

46 Miller-Urey Apparatus

Teaching Strategies

- Use this transparency when discussing the apparatus Stanley Miller used to simulate the conditions of a young Earth. Be sure students have an understanding of A. I. Oparin's primordial soup model.

- Have students describe the contents of Earth's early atmosphere as proposed by Oparin and Harold Urey. *(It lacked oxygen but was rich in nitrogen and hydrogen-containing gases.)*

How did Stanley Miller use these gases in his model, and what information did his model yield?

(Miller exposed the gases to electrical sparks, which simulated lightning. This produced a mixture of chemicals including life's basic building blocks: amino acids, fatty acids, and hydrocarbons.)

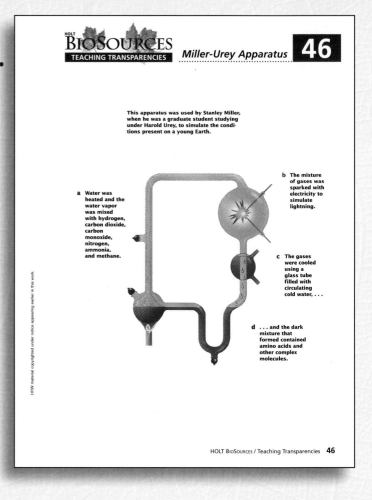

This apparatus was used by Stanley Miller, when he was a graduate student studying under Harold Urey, to simulate the conditions present on a young Earth.

a Water was heated and the water vapor was mixed with hydrogen, carbon dioxide, carbon monoxide, nitrogen, ammonia, and methane.

b The mixture of gases was sparked with electricity to simulate lightning.

c The gases were cooled using a glass tube filled with circulating cold water, . . .

d . . . and the dark mixture that formed contained amino acids and other complex molecules.

47 Spontaneous Assembly of RNA

Teaching Strategies

- Use this transparency when discussing the origin of nucleotides. Review the terms *nucleotide, RNA, adenine, guanine, cytosine,* and *uracil*.

- Walk students through the four steps of the illustration.

What kind of change in the environment could have caused the first chains of nucleotides to form?

(the removal of water, such as evaporation)

- In small groups, have students write a paragraph that answers the following question: Which came first, nucleic acids or proteins?

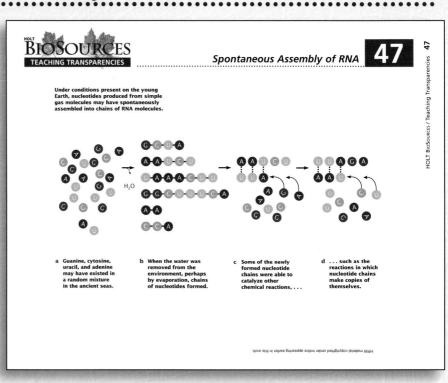

Under conditions present on the young Earth, nucleotides produced from simple gas molecules may have spontaneously assembled into chains of RNA molecules.

a Guanine, cytosine, uracil, and adenine may have existed in a random mixture in the ancient seas.

b When the water was removed from the environment, perhaps by evaporation, chains of nucleotides formed.

c Some of the newly formed nucleotide chains were able to catalyze other chemical reactions, . . .

d . . . such as the reactions in which nucleotide chains make copies of themselves.

BIOSOURCES
TEACHING TRANSPARENCIES

Evolution of the Horse **48**

The evolution of the modern horse began with *Hyracotherium* about 60 million years ago. Notice the change from four toes to one toe on each front foot.

a *Hyracotherium*
60 million
years ago

b *Mesohippus*
About 30 million
years ago

c *Merychippus*
About 20 million years ago

d *Pliohippus*
14 million to 7 million years ago

e *Equus* (modern horse)
10,000 years ago

HRW material copyrighted under notice appearing earlier in this work.

Teaching Strategies

- Use this transparency when discussing the evolution of the modern horse. Review the terms *species, transitional form, ancestor,* and *evolution.* Read the name of each transitional form aloud so that students can learn the correct pronunciations.
- Introduce students to the ancestors of the modern horse, emphasizing the period of time each existed. Ask students how long the modern horse has been evolving *(for over 60 million years).*

Which traits have changed significantly during the course of time? *(Four toes have changed to one toe on each front foot.)*

- Have students list other similarities and differences among these horses.

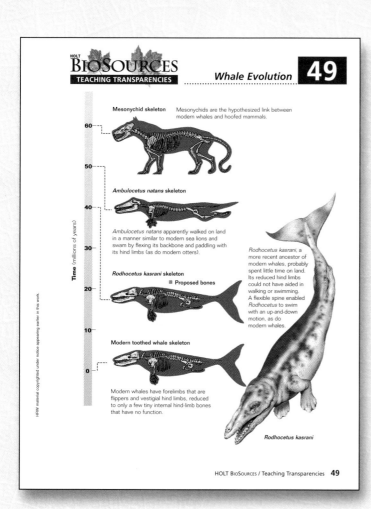

BIOSOURCES
TEACHING TRANSPARENCIES

Whale Evolution **49**

Mesonychid skeleton Mesonychids are the hypothesized link between modern whales and hoofed mammals.

Ambulocetus natans skeleton

Ambulocetus natans apparently walked on land in a manner similar to modern sea lions and swam by flexing its backbone and paddling with its hind limbs (as do modern otters).

Rodhocetus kasrani skeleton
■ Proposed bones

Rodhocetus kasrani, a more recent ancestor of modern whales, probably spent little time on land. Its reduced hind limbs could not have aided in walking or swimming. A flexible spine enabled *Rodhocetus* to swim with an up-and-down motion, as do modern whales.

Modern toothed whale skeleton

Modern whales have forelimbs that are flippers and vestigial hind limbs, reduced to only a few tiny internal hind-limb bones that have no function.

Time (millions of years)

60
50
40
30
20
10
0

HRW material copyrighted under notice appearing earlier in this work.

Rodhocetus kasrani

Whale Evolution **49**

Teaching Strategies

- Use this transparency when discussing the evolution of the modern whale. Read the names of each transitional form so that students can learn the correct pronunciations.
- Have volunteers describe similarities and differences in the bones of these animals. Point out that the backbone of the *Rodhocetus kasrani* skeleton is not complete. Explain that it is rare to find a complete skeleton of any one fossil. Emphasize that paleontologists use knowledge of anatomy to project the appearance of these bones.
- Ask students how the backbone of these animals changed in relation to the time the animals spent in water. *(The backbone became heavier and more flexible the longer the time the animal spent in water.)*

Teaching Strategies

• Use this transparency when discussing homologous structures. Review the term *homologous structure*, and read aloud the parts of the vertebrate forelimb so that students learn the correct pronunciations.

• Have students describe the similarities in the structures of these forelimbs. Point out that each vertebrate has the same complement of bones.

• Ask students why the individual bones of these limbs differ. *(Each has been modified to perform a different function.)*

Are the wings of birds and insects homologous structures? *(no)* **Why?** *(They have the same function but do not have the same structure.)*

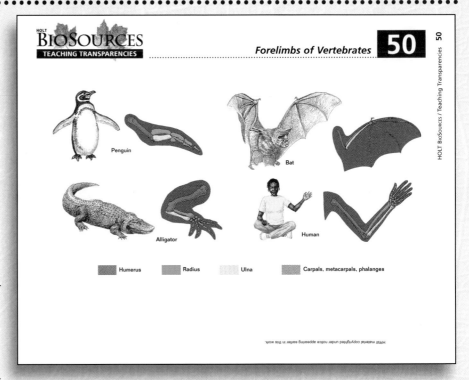

HOLT BIOSOURCES / TEACHING TRANSPARENCIES

Forelimbs of Vertebrates **50**

Penguin

Bat

Alligator

Human

Humerus · Radius · Ulna · Carpals, metacarpals, phalanges

HOLT BioSources / Teaching Transparencies 50

Teaching Strategies

• Use this transparency when comparing the two hypotheses on evolutionary rates. Review the terms *gradualism* and *punctuated equilibrium*.

• Walk students through this diagram.

How does gradualism account for the gaps in the fossil record?

(The gaps in the fossil record are a weakness in the theory of gradualism, as Darwin acknowledged in his writings.)

What about punctuated equilibrium?

(According to punctuated equilibrium, gaps in the fossil record are expected.)

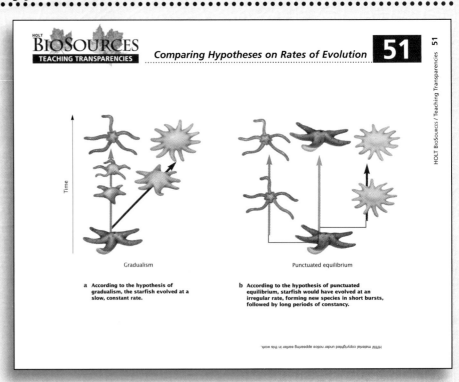

HOLT BIOSOURCES / TEACHING TRANSPARENCIES

Comparing Hypotheses on Rates of Evolution **51**

Time

Gradualism

Punctuated equilibrium

a According to the hypothesis of gradualism, the starfish evolved at a slow, constant rate.

b According to the hypothesis of punctuated equilibrium, starfish would have evolved at an irregular rate, forming new species in short bursts, followed by long periods of constancy.

HOLT BioSources / Teaching Transparencies 51

BioSOURCES
TEACHING TRANSPARENCIES

Process of
Natural Selection 52

Steps	Explanation	Example
Variation is the raw material for natural selection.	Every species contains genetic variation: individuals differ because they carry different alleles for a certain trait. Mutation, sexual reproduction, and crossing over produce unique combinations of alleles.	Giraffes were born with alleles for varying neck lengths: short and long.
Living things face a constant struggle for existence.	Organisms produce more offspring than can survive. These offspring must evade predators and compete for food and living space.	Giraffes with longer necks could reach the leaves in tall trees.
Only some individuals survive and reproduce.	Some individuals survive better than others. These individuals are more likely to produce offspring.	The giraffes with longer necks were better at getting food. Consequently, they produced more offspring than did giraffes with short necks.
Natural selection causes genetic change.	Each generation consists of offspring of successful individuals. Thus, it contains an increased proportion of individuals with traits that promote survival and reproduction than did the previous generation.	Over time, long-necked giraffes became common and replaced short-necked giraffes.
Species adapt to their environment.	Selection tends to make a population better suited to its environment. The environment determines which alleles are favored.	Long-necked giraffes are able to reach the leaves on tall trees, which are out of reach for most other animals.

HOLT BioSOURCES / Teaching Transparencies 52

Teaching Strategies

- Use this transparency when discussing the general steps involved in the process of natural selection. Be sure that students are familiar with the terms *genetic variation, allele, mutation, sexual reproduction,* and *crossing over.*

- Walk students through the steps in the process of natural selection, as shown in this illustration. Have them describe the characteristics of a "successful individual" as shown in the example in the table. *(A successful individual is one that has a trait that enables it to perform a task of life better than other individuals can. For example, giraffes with long necks were better at getting food, and they consequently produced more offspring than giraffes with short necks.)*

BioSOURCES
TEACHING TRANSPARENCIES

Evolutionary
Relationships of
Anthropoids 53

Gibbons Orangutan Gorilla Bonobo Chimpanzee Humans

Millions of years ago

This evolutionary tree shows the close relationship between humans and apes. Going back in time, . . .

a . . . humans shared a common ancestor with chimpanzees and bonobos until about 5 million years ago.

b About 7 million years ago the gorilla's ancestor diverged from the ancestor of chimpanzees, bonobos, and humans.

c The ancestor of orangutans diverged about 8 million years ago. The ancestors of gibbons diverged from the other apes earlier, about 10 million years ago.

HOLT BioSOURCES / Teaching Transparencies 53

Teaching Strategies

- Use this transparency when discussing the evolutionary relationships of anthropoids. Be sure students are aware that anthropoids are day-active primates—monkeys, apes, and humans. Have students design a graphic organizer that illustrates the characteristics of monkeys, apes, and humans.

- Emphasize that this evolutionary tree shows that apes and humans share a common ancestor, not that humans descended from apes.

How many years ago did the common ancestor of gorillas, chimpanzees, bonobos, and humans exist?

(about 5 million years ago)

- Inform students that biochemical evidence suggests that humans are most closely related to the chimpanzees, less closely related to the gorillas, even more distantly related to the orangutans, and most distantly related to the gibbons.

Comparison of Chimpanzee and Human Jaws

Teaching Strategies

- Use this transparency when comparing hominid and chimpanzee jaws. Be sure students are familiar with the four types of teeth found in both kinds of jaw.
- Have students imagine that they are fossil hunters in Africa and that they have just uncovered the remains of a jaw with six molars and four premolars in tact.

How would you determine whether the jaw belonged to a hominid or to a chimpanzee?

(Examine the arrangement of the molars. If they are arranged in two parallel rows, the jaw belonged to a chimpanzee; if they are not parallel, it belonged to a hominid.)

- Have students hypothesize about how chimpanzees use their relatively long canines.

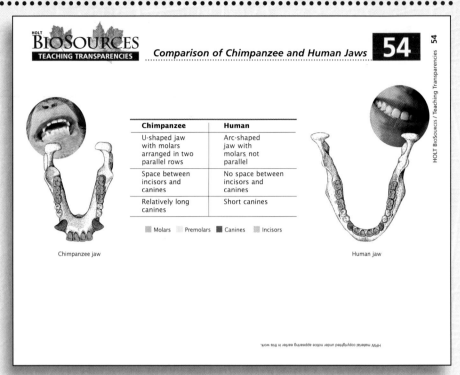

HOLT BIOSOURCES — TEACHING TRANSPARENCIES

Comparison of Chimpanzee and Human Jaws **54**

Chimpanzee	Human
U-shaped jaw with molars arranged in two parallel rows	Arc-shaped jaw with molars not parallel
Space between incisors and canines	No space between incisors and canines
Relatively long canines	Short canines

Molars Premolars Canines Incisors

Chimpanzee jaw   Human jaw

Comparison of Gorilla and Australopithecine Skeletons

Teaching Strategies

- Use this transparency when comparing the skeletons of gorillas and australopithecines. Be sure students are familiar with the following parts of the skeleton: skull, spine, arms, pelvis, and femur.
- Have students imagine that they are fossil hunters and that they have uncovered the remains of a skull and spinal column.

How will you determine whether the remains belonged to a gorilla or to an australopithecine?

(Examine how the spinal cord exits the skull.)

- Have students hypothesize about why gorillas have arms that are longer than their legs.

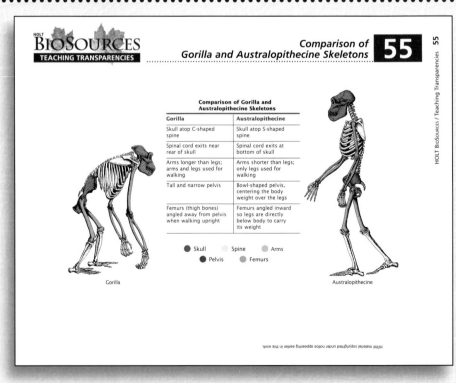

HOLT BIOSOURCES — TEACHING TRANSPARENCIES

Comparison of Gorilla and Australopithecine Skeletons **55**

Comparison of Gorilla and Australopithecine Skeletons

Gorilla	Australopithecine
Skull atop C-shaped spine	Skull atop S-shaped spine
Spinal cord exits near rear of skull	Spinal cord exits at bottom of skull
Arms longer than legs; arms and legs used for walking	Arms shorter than legs; only legs used for walking
Tall and narrow pelvis	Bowl-shaped pelvis, centering the body weight over the legs
Femurs (thigh bones) angled away from pelvis when walking upright	Femurs angled inward so legs are directly below body to carry its weight

Skull Spine Arms Pelvis Femurs

Gorilla   Australopithecine

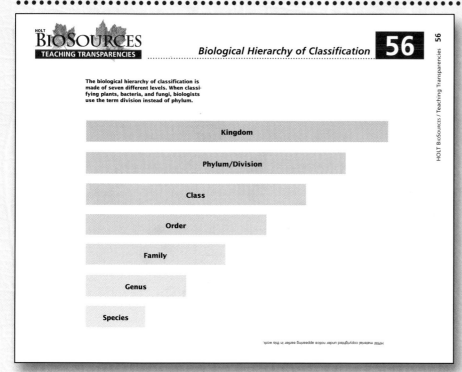

BioSOURCES
TEACHING TRANSPARENCIES

Biological Hierarchy of Classification 56

The biological hierarchy of classification is made of seven different levels. When classifying plants, bacteria, and fungi, biologists use the term division instead of phylum.

- Kingdom
- Phylum/Division
- Class
- Order
- Family
- Genus
- Species

Teaching Strategies

- Use this transparency when explaining the biological hierarchy of classification. Be sure students understand that each category in this hierarchy is a collective unit containing one or more groups from the next-lower level in the hierarchy. Thus, a genus is a group of closely related species, a family is a group of closely related genera, and so on.
- Emphasize that this taxonomic hierarchy classifies living things on the basis of phylogenetic relationships.
- Have students explore the history of the modern system of classification, particularly Linnaeus's contribution to the naming of species.

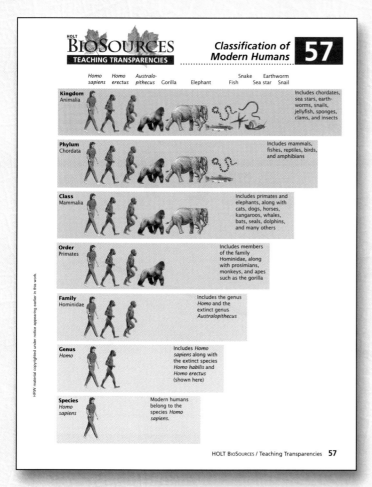

BioSOURCES
TEACHING TRANSPARENCIES

Classification of Modern Humans 57

	Homo sapiens	Homo erectus	Australo-pithecus	Gorilla	Elephant	Fish	Snake Sea star	Earthworm Snail	

Kingdom Animalia — Includes chordates, sea stars, earthworms, snails, jellyfish, sponges, clams, and insects

Phylum Chordata — Includes mammals, fishes, reptiles, birds, and amphibians

Class Mammalia — Includes primates and elephants, along with cats, dogs, horses, kangaroos, whales, bats, seals, dolphins, and many others

Order Primates — Includes members of the family Hominidae, along with prosimians, monkeys, and apes such as the gorilla

Family Hominidae — Includes the genus *Homo* and the extinct genus *Australopithecus*

Genus Homo — Includes *Homo sapiens* along with the extinct species *Homo habilis* and *Homo erectus* (shown here)

Species Homo sapiens — Modern humans belong to the species *Homo sapiens*.

Teaching Strategies

- Use this transparency when discussing the taxonomical classification of modern humans. Review the seven levels of taxa in the biological hierarchy.
- Lead students through the classification of modern humans. Ask students in which taxon humans are classified with fishes, reptiles, birds, and amphibians *(in the phylum Chordata)*.

In which taxon are humans classified with other primates only?
(in the order Primates)

Why aren't plants included in this classification table?
(Plants belong to the kingdom Plantae.)

- In small groups, have students select a favorite organism, and have them design a similar table describing its classification. Encourage them to choose organisms from the kingdom Plantae as well.

Teaching Strategies

- Use this transparency when discussing methods of taxonomy. Review the terms *cladistics, derived characters,* and *cladogram.*

- Walk students through the cladogram of the major divisions of plants. Ask students how mosses differ from ferns, pine trees, and flowering plants. *(Ferns lack vascular tissues.)*

How do mosses, ferns, and pine trees differ from flowering plants?

(They do not produce flowers.)

- Emphasize to students that cladists classify groups based on shared derived characteristics. Cladistic methodology dominates systematics today.

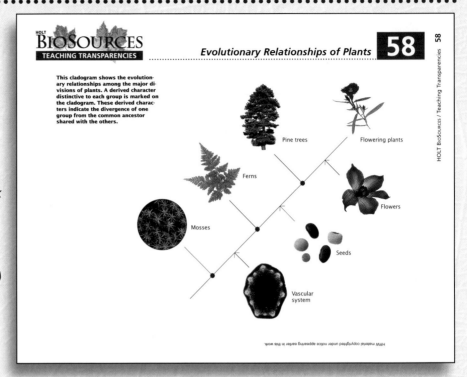

Teaching Strategies

- Use this transparency as an example of a cladogram of seven familiar vertebrates. Be sure that students are aware that derived traits listed along the horizontal axis are shared by all of the organisms to the right of the branch points and are not present in any organisms to the left.

- Walk students through the cladogram of the seven vertebrates. Ask them how lampreys differ from sharks, salamanders, lizards, tigers, gorillas, and humans. *(Lampreys lack jaws.)*

- Have students hypothesize about the derived trait that could be listed to distinguish gorillas from humans.

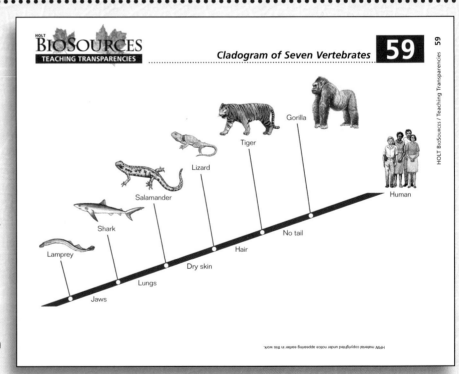

BIOSOURCES
TEACHING TRANSPARENCIES

Bird Phylogeny and
DNA Sequencing
60

Millions of years ago

50   40   30   20   10   0

Comparison of the DNA sequences of these birds reveals their evolutionary relationships. For example, the DNA sequences of flamingos differ by the greatest amount. Their ancestor diverged from the other birds about 50 million years ago.

Flamingos

Ibises

Shoebills

Pelicans

Storks

New World vultures

HRW material copyrighted under notice appearing earlier in this work.

Bird Phylogeny and DNA Sequencing 60

Teaching Strategies

- Use this transparency when discussing how taxonomists use technology to establish evolutionary relationships in a group of organisms. Review the terms *phylogeny, DNA, nucleotide sequence*, and *mutation*.

- Walk students through the evolutionary tree of the six birds shown in the illustration. Ask students what the tree reveals about pelicans and shoebills. *(They have similar DNA sequences and evolved from a common ancestor about 37 million years ago.)*

What does the tree say about ibises, shoebills, pelican, storks, and New World vultures?

(They evolved from a common ancestor about 47 million years ago.)

BIOSOURCES
TEACHING TRANSPARENCIES

Theory of
Endosymbiosis
61

Larger host cell

Nucleus

Aerobic eubacterium

Mitochondrion

Eukaryotic cell

According to the theory of endosymbiosis, eukaryotic cells evolved when aerobic eubacteria either infected or were engulfed by a larger host cell and later established a symbiotic relationship.

HRW material copyrighted under notice appearing earlier in this work.

Theory of Endosymbiosis 61

Teaching Strategies

- Use this transparency when explaining the theory of endosymbiosis. Review the terms *symbiosis, prokaryotic cell, eukaryotic cell, aerobic eubacterium*, and *mitochondrion*.

- Guide students through the events of endosymbiosis. Have them identify the host cell and the aerobic eubacterium. Explain that according to the theory, these two organisms established a symbiotic relationship and evolved into the eukaryotic cell. Ask students to describe a process that would enable the host cell to capture the bacterium. *(Students should describe endocytosis.)*

- Have students design an illustration that shows the evolution of a chloroplast, using the transparency as a model if needed.

62 Two Hypotheses on the Evolution of Hominids

Teaching Strategies

- Use this transparency when discussing the evolution of hominids. Review the term *hominid*.
- Walk students through the two hypotheses illustrated on the transparency. Have students explain the differences between the two hypotheses. *(The main differences lie in the evolution of* A. robustus *and the time at which the three lines of hominids split from* A. afarensis.*)*

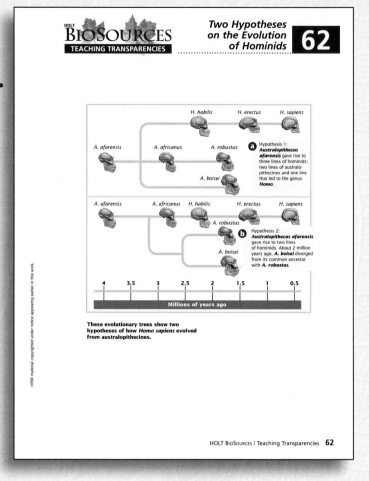

HOLT
BioSources
TEACHING TRANSPARENCIES

Two Hypotheses
on the Evolution
of Hominids 62

These evolutionary trees show two hypotheses of how *Homo sapiens* evolved from australopithecines.

63 Stabilizing, Directional, and Disruptive Selection

Unit 4 *Ecology*

Teaching Strategies

- Use this transparency when discussing the three types of selection pressures involved in adaptation. Review the terms *natural selection* and *population*.
- Walk students through the graphs. Explain that each type of selection pressure has a different effects on the appearance of a trait in a population. Have students design a graphic organizer comparing the three types of selection pressures and including examples.

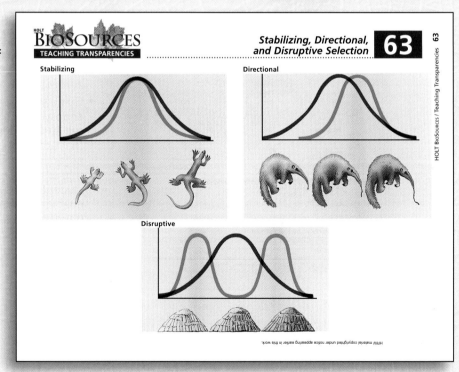

Stabilizing

Directional

Disruptive

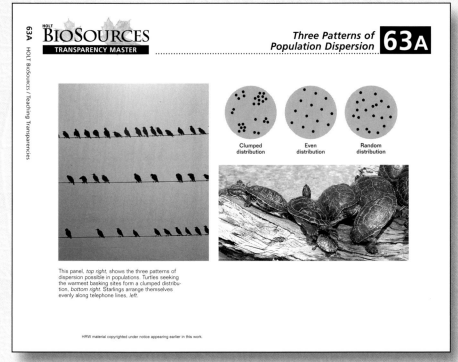

HOLT BioSources
TRANSPARENCY MASTER

Three Patterns of Population Dispersion **63A**

Clumped distribution   Even distribution   Random distribution

This panel, *top right*, shows the three patterns of dispersion possible in populations. Turtles seeking the warmest basking sites form a clumped distribution, *bottom right*. Starlings arrange themselves evenly along telephone lines, *left*.

HRW material copyrighted under notice appearing earlier in this work.

Teaching Strategies

- Use this transparency when discussing the patterns of population dispersion. Review the terms *population* and *ecosystem*.

- Discuss the three patterns of population dispersion. Point out that the dispersion pattern for a particular population depends on the scale of observation. Give an example.

- Have students hypothesize about the pattern of dispersion for each of the following populations: grass plants in which individual plants inhibit seedling germination in a circular zone around their base *(even)*; a flock of flamingos *(clumped)*; and mushrooms growing on a forest floor *(random)*.

Map of Homo sapiens Migration 64

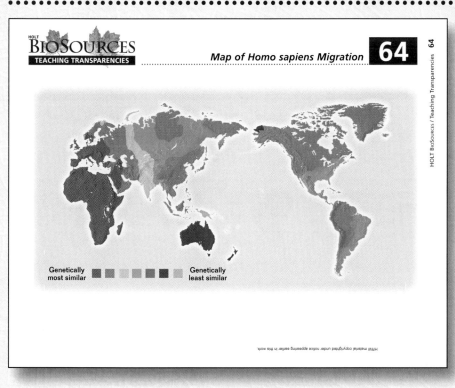

HOLT BioSources
TEACHING TRANSPARENCIES

Map of Homo sapiens Migration **64**

Genetically most similar   Genetically least similar

HRW material copyrighted under notice appearing earlier in this work.

Teaching Strategies

- Use this transparency to discuss the migration of *Homo sapiens* from Africa throughout the world. Review the significance of the terms *mitochondrial DNA*, *Homo erectus*, and *Neanderthal*.

- Review the species' path of migration. Have students locate the continent on which the highest degree of genetic similarity can be found *(Africa)*.

- Have students hypothesize why there are different varieties of *H. sapiens*. *(The unique characteristics of each human variety represent adaptations to local conditions that evolved after our species migrated from Africa.)*

Teaching Strategies

- Use this transparency when discussing the two strategies of population growth. Review the terms *carrying capacity, r-strategist,* and *k-strategist.*
- Inform students that exponential growth begins very slowly and accelerates as the number of reproducing individuals increases. Logistic growth slows as resources become scarce and wastes accumulate.
- Ask students to identify the three main stages of logistic growth: (1) establishment of the population *(the far left portion of the curve),* (2) exponential growth *(the lower half of the curve),* and (3) stabilization around the carrying capacity *(the far right portion of the curve).*

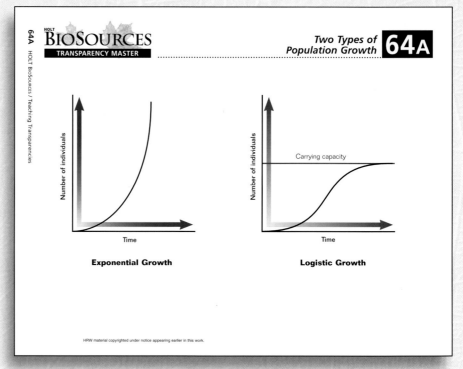

HOLT
BIOSOURCES
TRANSPARENCY MASTER

Two Types of
Population Growth **64A**

Exponential Growth

Logistic Growth

Carrying capacity

Number of individuals

Number of individuals

Time

Time

64A   HOLT BioSources / Teaching Transparencies

HRW material copyrighted under notice appearing earlier in this work.

Teaching Strategies

- Use this transparency when discussing the global distribution of biomes. Review the terms *biome, tropical, temperate,* and *deciduous.*
- Emphasize that a biome is a category, not a place. A tropical rain forest refers not to any specific geographical location on Earth but to all regions of the planet where such a biome can be found. Point out that biomes are characterized by their dominant vegetation.
- Have students make a table showing the types and locations of biomes on each continent.

Which biomes are not found in the United States?

(savanna and tropical rain forest)

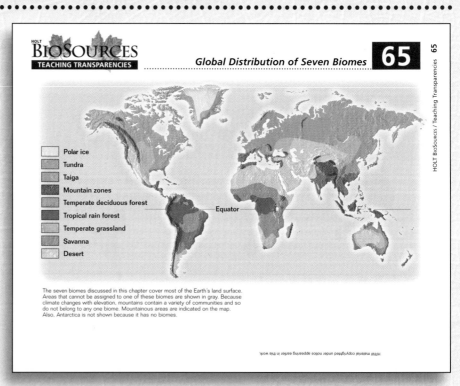

HOLT
BIOSOURCES
TEACHING TRANSPARENCIES

Global Distribution of Seven Biomes **65**

- Polar ice
- Tundra
- Taiga
- Mountain zones
- Temperate deciduous forest
- Tropical rain forest
- Temperate grassland
- Savanna
- Desert

Equator

The seven biomes discussed in this chapter cover most of the Earth's land surface. Areas that cannot be assigned to one of these biomes are shown in gray. Because climate changes with elevation, mountains contain a variety of communities and so do not belong to any one biome. Mountainous areas are indicated on the map. Also, Antarctica is not shown because it has no biomes.

HRW material copyrighted under notice appearing earlier in this work.

65   HOLT BioSources / Teaching Transparencies

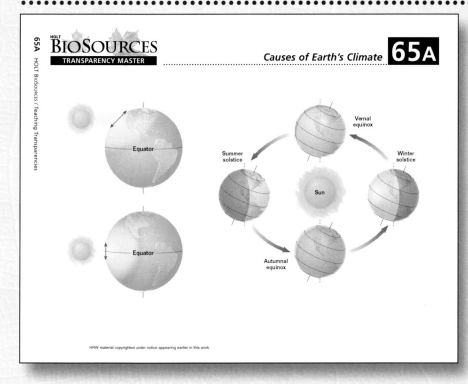

HOLT
BIOSOURCES
TRANSPARENCY MASTER

Causes of Earth's Climate 65A

65A

HOLT BioSources / Teaching Transparencies

Equator

Vernal equinox

Summer solstice

Winter solstice

Sun

Autumnal equinox

Equator

HRW material copyrighted under notice appearing earlier in this work.

Teaching Strategies

- Use this transparency when discussing the factors that determine climate. Review the terms *latitude* and *hemisphere*.

- Inform students that latitude is the most important factor influencing climate. Latitude determines the angle at which solar energy strikes the Earth's surface. Areas that receive direct solar radiation are warmer than areas that receive diffuse solar radiation.

- Have students explain why it is winter in the Southern Hemisphere when it is summer in the Northern Hemisphere. *(When one hemisphere tilts toward the sun, the other tilts away, causing opposite seasons in the two hemispheres.)*

Ecological Races of Seaside Sparrows 66

HOLT
BIOSOURCES
TEACHING TRANSPARENCIES

Ecological Races of Seaside Sparrows 66

66

HOLT BioSources / Teaching Transparencies

Ecological Range and Distribution of the Seaside Sparrow

Seaside sparrow
Gulf seaside sparrow
Dusky seaside sparrow
Cape Sable seaside sparrow

Seaside sparrow
(Ammodramus maritimus maritimus)
Widely distributed along the Atlantic and Gulf coasts from Massachusetts to southern Texas, this sparrow is noted for making "crash landings" into marsh grasses.

Dusky seaside sparrow
(Ammodramus maritimus nigrescens)
Now extinct, this race occurred only near Merritt Island and in the marshes of the St. John's River near Titusville, Florida.

Cape Sable seaside sparrow
(Ammodramus maritimus mirabilis)
Found only in Cape Sable on the southwestern tip of Florida, this very local ecological race is the only seaside sparrow living in the area.

Gulf seaside sparrow
(Ammodramus maritimus fisheri)
Found along the Gulf Coast, this sparrow's range overlaps the range of the more common

HRW material copyrighted under notice appearing earlier in this work.

Teaching Strategies

- Use this transparency when discussing ecological races. Review the terms *divergence, speciation*, and *natural selection*.

- Have students identify the genus and species of each seaside sparrow. Tell students that microevolution can result in the development of ecological races.

- For each race of seaside sparrow, have students write a sentence that describes its range.

- Ask students what each sparrow's third name indicates about the relationship of these populations. *(They are different ecological races.)*

When will two races of seaside sparrows become different species? *(when they can no longer interbreed successfully)*

Teaching Strategies

- Use this transparency to display an example of a food web. Review the terms *producer, consumer*, and *trophic level*.

- Trace the paths of energy through the grassland food web. Have students identify the organisms in each trophic level, beginning with the producers.

Which organism falls into the class of consumers called detritivores?

(mushrooms)

- Ask students to imagine how the food web would be affected by the following situations: a series of wildfires, a prolonged drought, a sudden flood, suburban development. Have them write a paragraph that describes each of these effects.

Grassland Food Web **67**

Decomposers

Teaching Strategies

- Use this transparency to display an example of a food chain. Remind students that a food chain is a single pathway of energy flow in an ecosystem. Also remind students of the importance of producers in ecosystems.

- Have students identify the producers, herbivores, and carnivores in this diagram. *(Algae are the producers; krill are the herbivores; and cod, leopard seals, and killer whales are carnivores.)*

- Ask students which organisms, if any, are omnivores or detritivores. *(There are no omnivores or detritivores.)*

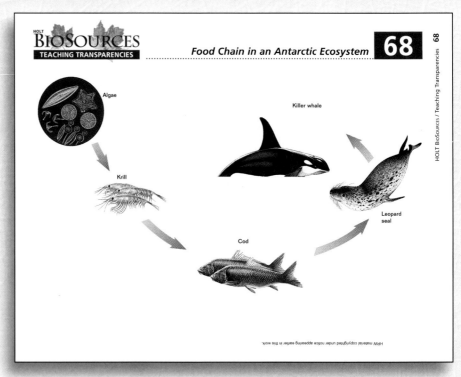

Food Chain in an Antarctic Ecosystem **68**

Algae

Killer whale

Krill

Cod

Leopard seal

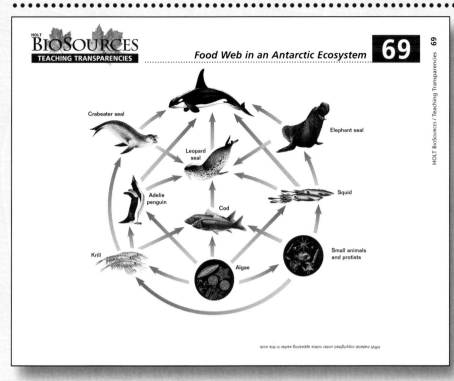

BIOSOURCES
TEACHING TRANSPARENCIES

Food Web in an Antarctic Ecosystem **69**

HOLT BioSources / Teaching Transparencies 69

Crabeater seal

Elephant seal

Leopard seal

Adelie penguin

Squid

Cod

Krill

Algae

Small animals and protists

HRW material copyrighted under notice appearing earlier in this work.

Teaching Strategies

- Use this transparency to display an example of a food web. Emphasize the interconnectedness of the ecosystem shown in the diagram. Because of the interconnectedness, a change in the numbers of one species can affect many other species in the ecosystem.

- Ask students how this ecosystem would be affected if the algae died out. *(Students should recognize that algae are the producers in this ecosystem and capture the energy needed by other members of the ecosystem. Without algae, the other inhabitants would eventually starve.)*

Ecological Succession at Glacier Bay 70

BIOSOURCES
TEACHING TRANSPARENCIES

Ecological Succession at Glacier Bay **70**

HOLT BioSources / Teaching Transparencies 70

At first, land exposed by the receding glacier is lifeless because it lacks nutrients. An early "pioneer" of this land is the rockrose *Dryas, above left.* After several decades trees such as alder and shrubs grow large enough to shade and kill off the low-growing mat of *Dryas, above center.* After several more decades, these trees and shrubs are replaced by spruce and hemlock, *above right.*

HRW material copyrighted under notice appearing earlier in this work.

Teaching Strategies

- Use this transparency when discussing ecological succession. Have students locate Glacier Bay, Alaska, on a world map. Review the terms *succession, primary succession*, and *secondary succession.*

- Inform students that succession is a process they have probably seen in action. Weeds, which are pioneer organisms, quickly invade any unoccupied habitat, including abandoned fields, freshly tilled soil, and even cracks in sidewalks and parking lots.

- Ask students what would happen if the spruce-hemlock climax community were destroyed by fire. *(Succession would occur again, beginning with the pioneer plants similar to those that first appeared at Glacier Bay.)*

Teaching Strategies

- Use this transparency when discussing the water cycle. Review the terms *precipitation, runoff, seepage*, and *evaporation*.

- Trace the path that a molecule of water follows through the water cycle, starting with a molecule in a raindrop that has just fallen on the ground. Point out that a molecule of water spends differing amounts of time in each component of the water cycle (atmosphere, ocean, lake, groundwater).

- Ask students to determine in which component a molecule of water remains the longest *(in ground water)*.

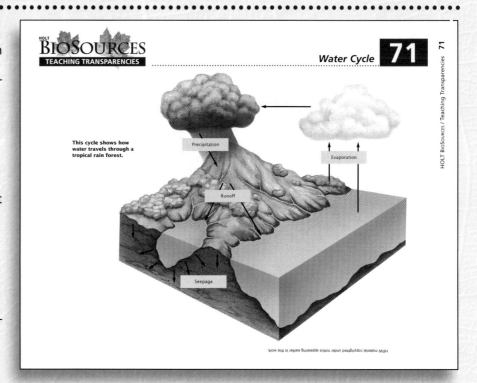

Teaching Strategies

- Use this transparency when discussing the carbon cycle. Review the terms *photosynthesis, cellular respiration*, and *decomposition*.

- Inform students that the burning of fossil fuels has released much more carbon dioxide into the atmosphere than can be recycled through photosynthesis or absorbed by the oceans. Because carbon dioxide helps give the Earth a moderate climate, scientists are concerned about the possibility of global warming caused by the excess carbon dioxide.

- Have students list three ways humans add carbon dioxide to the atmosphere *(exhaling carbon dioxide, burning fossil fuels and wood, cutting vegetation that decomposes)*.

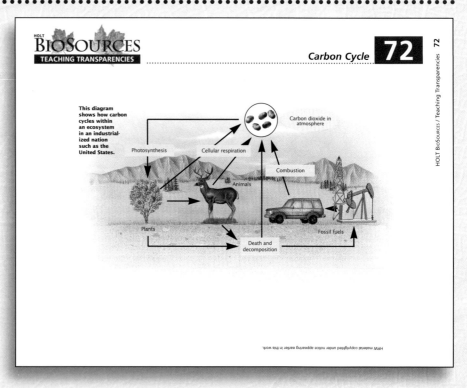

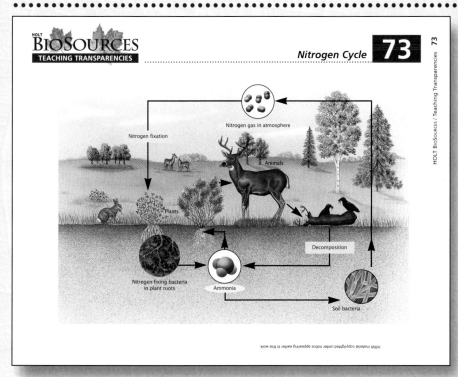

Teaching Strategies

- Use this transparency when discussing the nitrogen cycle. Review the terms *nitrogen fixation, ammonification, nitrification,* and *denitrification.*

- Point out the difference between nitrogen fixation and nitrification. Tell students that lightning also results in nitrogen fixation, although such atmospheric action amounts to less than 10 percent of nitrogen fixation by organisms.

- Have students explain why nitrogen is often scarce in the soil even though the atmosphere is 78 percent nitrogen gas. *(Only a few kind of bacteria can use nitrogen gas.)*

Carbon Dioxide and Average World Temperatures 74

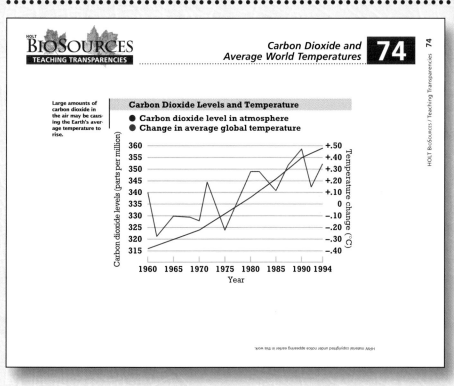

Teaching Strategies

- Use this transparency when discussing the increase in carbon dioxide levels and average world temperatures. Review the terms *greenhouse effect* and *global warming.*

- Emphasize that this graph shows the correlation between atmospheric carbon dioxide concentrations and average temperature for the past thirty years.

- Inform students that of the carbon dioxide released by the burning of fossil fuel, about half remains in the atmosphere. The rest is absorbed by the ocean waters. Add that the carbon dioxide reservoir in the atmosphere is being increased through deforestation, particularly in the tropics, which is an important "sink" for the gas.

Teaching Strategies

- Use this transparency to illustrate the dramatic growth of the world's human population over the past fifty years. Emphasize that the human population has slowly increased over thousands of years.

- Review the term *carrying capacity*. Inform students that this concept is as valid for the human population as it is for other organisms. In the next century, the Earth will have to support twice as many humans as it does today. Have students make a list of issues that will arise from this increase, and use the list as a basis for discussion.

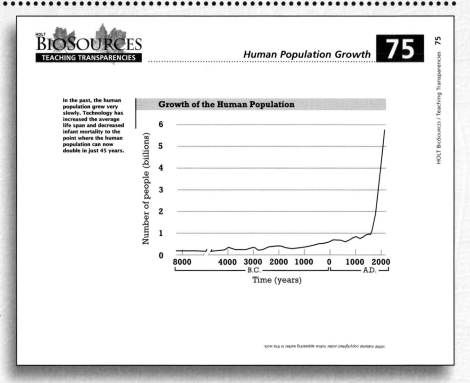

Teaching Strategies

- Use this transparency when discussing the projected human population growth. Point out that the orange rectangles represent the current population and that the purple rectangles represent the projected population in 2025.

- Review the current and estimated populations for each country in the diagram. Contrast the rapid growth projected for developing countries with the slow growth predicted for industrialized countries.

- Have students calculate the expected increase in the populations of the United States, India, and Kenya, expressed as a percentage *(United States—30 percent; India—51 percent; Kenya—137 percent)*.

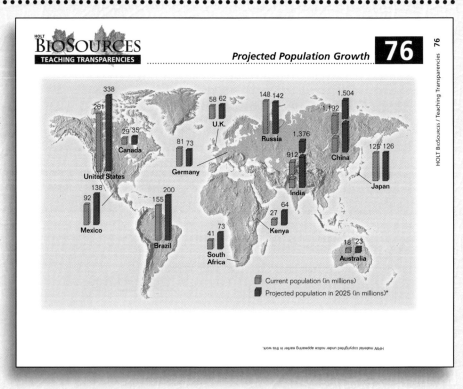

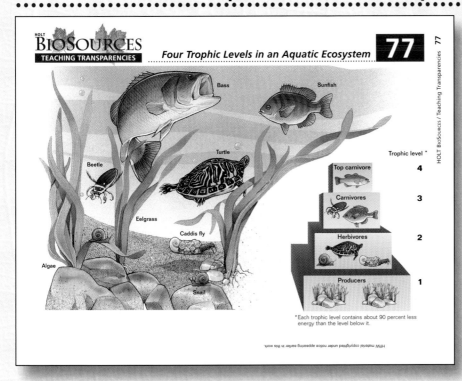

Teaching Strategies

- Use this transparency when discussing aquatic ecosystems. Review the terms *trophic level, producers, herbivore*, and *carnivore*.

- Remind students of the second law of thermodynamics and its consequences for energy transfers within ecosystems.

- Have students consider how the efficiency of the predators in this ecosystem might affect the pyramid of energy. For example, if bass are particularly inefficient at capturing prey, how will that affect the size of the fourth trophic level? *(The less efficient the bass are at capturing prey, the smaller their trophic level will be.)*

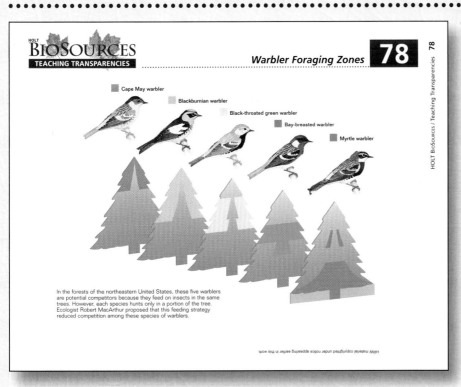

In the forests of the northeastern United States, these five warblers are potential competitors because they feed on insects in the same trees. However, each species hunts only in a portion of the tree. Ecologist Robert MacArthur proposed that this feeding strategy reduced competition among these species of warblers.

Teaching Strategies

- Use this transparency when discussing examples of a species' niche.

- Inform students that ecologists often use the term *resource partitioning* to describe the patterns of resource use in a community. Ask them to explain why resource partitioning is an appropriate description of the feeding behaviors of the five warbler species studied by MacArthur.

- Competition can cause a species to alter its use of resources. Ask students to explain how predation might affect the niche of a prey species. *(The prey may change its habits to avoid the predator. For example, it might feed only at night.)*

Teaching Strategies

- Use this transparency when discussing marine communities. Review the process of photosynthesis and its significance in biological systems.

- Have students identify the three zones in the ocean: (1) the shallow coastal zone; (2) the open ocean surface; and (3) the depths of the open ocean. Point out that the characteristics of each zone are determined largely by the amount of solar energy and nutrients the zone receives. For instance, the shallow coastal zone is very productive because it receives nutrients from land.

- Ask students to explain why photosynthesis can't occur in the depths of the ocean. *(Not enough light can penetrate to these depths.)*

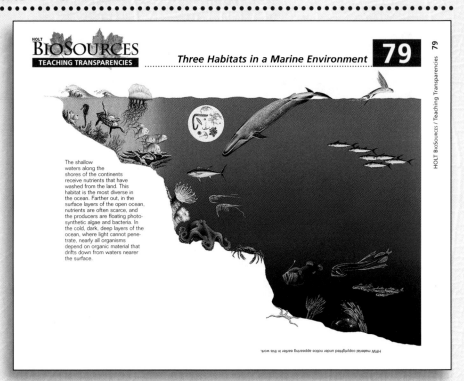

The shallow waters along the shores of the continents receive nutrients that have washed from the land. This habitat is the most diverse in the ocean. Farther out, in the surface layers of the open ocean, nutrients are often scarce, and the producers are floating photosynthetic algae and bacteria. In the cold, dark, deep layers of the ocean, where light cannot penetrate, nearly all organisms depend on organic material that drifts down from waters nearer the surface.

80 *Succession in a Developing Ecosystem*

Teaching Strategies

- Use this transparency when discussing succession in a developing ecosystem. Review the terms *succession, pioneer community, primary succession*, and *secondary succession*.

- Have students form a hypothesis about the types of plants that would be involved in primary succession on the slopes of Mount St. Helens *(weeds, grasses, mosses, ferns)*.

What types of plants would be involved in secondary succession?

(shrubs and tree seedlings, giving way to pine forests)

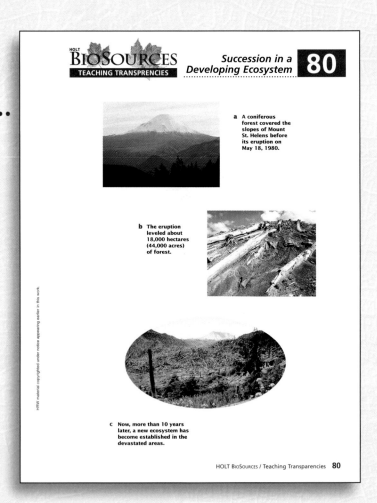

a A coniferous forest covered the slopes of Mount St. Helens before its eruption on May 18, 1980.

b The eruption leveled about 18,000 hectares (44,000 acres) of forest.

c Now, more than 10 years later, a new ecosystem has become established in the devastated areas.

HOLT BioSources / Teaching Transparencies **80**

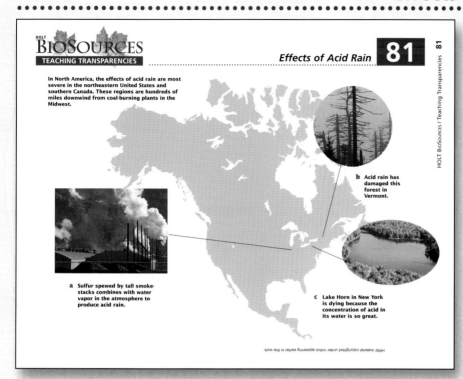

BIOSOURCES
TEACHING TRANSPARENCIES

Effects of Acid Rain 81

HOLT BioSources / Teaching Transparencies 81

In North America, the effects of acid rain are most severe in the northeastern United States and southern Canada. These regions are hundreds of miles downwind from coal-burning plants in the Midwest.

b Acid rain has damaged this forest in Vermont.

a Sulfur spewed by tall smoke-stacks combines with water vapor in the atmosphere to produce acid rain.

c Lake Horn in New York is dying because the concentration of acid in its water is so great.

Teaching Strategies

• Use this transparency when discussing the devastating effects of acid rain. Review the terms acid and pH.

• Inform students that living cells are extraordinarily sensitive to pH. Most cells function best when their interior is at a near-neutral pH of 6.8

What combination of chemicals produces acid rain?

(sulfur and water vapor)

What factors contribute to the severity of acid rain in the northeastern United States and Canada?

(winds and the burning of fossil fuels upwind from these regions)

Making an Ecosystem Model 82

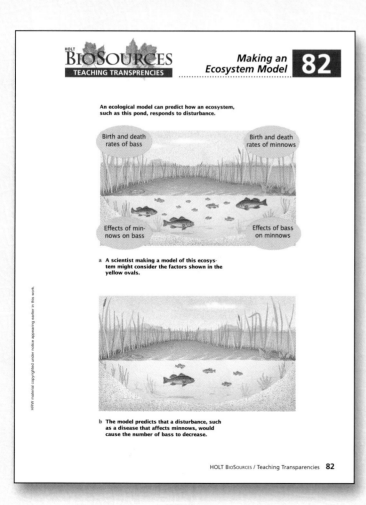

BIOSOURCES
TEACHING TRANSPRENCIES

Making an Ecosystem Model 82

An ecological model can predict how an ecosystem, such as this pond, responds to disturbance.

Birth and death rates of bass

Birth and death rates of minnows

Effects of minnows on bass

Effects of bass on minnows

a A scientist making a model of this ecosystem might consider the factors shown in the yellow ovals.

b The model predicts that a disturbance, such as a disease that affects minnows, would cause the number of bass to decrease.

Teaching Strategies

• Use this transparency when discussing how ecologists study ecosystems. Review the terms *community, habitat*, and *diversity*.

• Walk students through the steps an ecologist follows to construct an ecological model. Emphasize that before ecologists—and all other scientists—can build a model, they must first ask questions.

Teaching Strategies

- Use this transparency when discussing the factors that affect ecosystem diversity.
- Emphasize that the diversity of an ecosystem is affected by two factors—the size of the ecosystem and its latitude. Point out that this table shows the percentage of large-mammal species lost in eight national parks since their founding. Ask students to explain why the founding of a national park would reduce the number of large-mammal species. *(National parks invite a degree of development and establish a fairly constant human presence, both of which deter populations of large mammals.)*

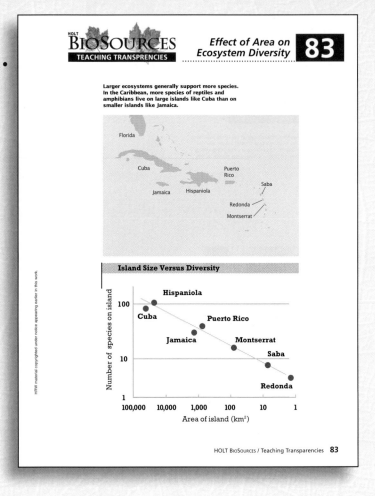

HOLT BioSOURCES / TEACHING TRANSPARENCIES

Effect of Area on Ecosystem Diversity **83**

Larger ecosystems generally support more species. In the Caribbean, more species of reptiles and amphibians live on large islands like Cuba than on smaller islands like Jamaica.

Island Size Versus Diversity

HOLT BioSOURCES / Teaching Transparencies **83**

84 *Energy Flow Through an Ecosystem*

Teaching Strategies

- Use this transparency when discussing energy flow through an ecosystem.
- Have students design a hypothetical food web that illustrates the flow of energy through an ecosystem that contains the following organisms: foxes, rabbits, plants, hawks, squirrels, owls, mice, snakes, seed-eating birds, insectivorous birds, spiders, herbivorous insects, predaceous insects, and toads.

Which class of organisms have been omitted from this web?

(decomposers)

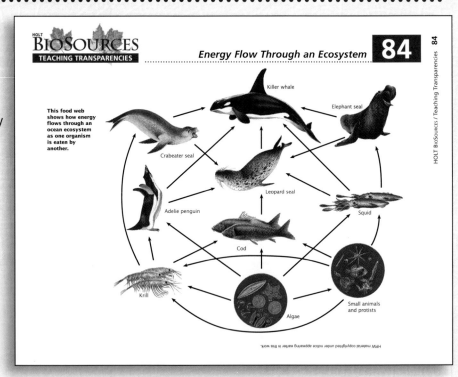

HOLT BioSOURCES / TEACHING TRANSPARENCIES

Energy Flow Through an Ecosystem **84**

This food web shows how energy flows through an ocean ecosystem as one organism is eaten by another.

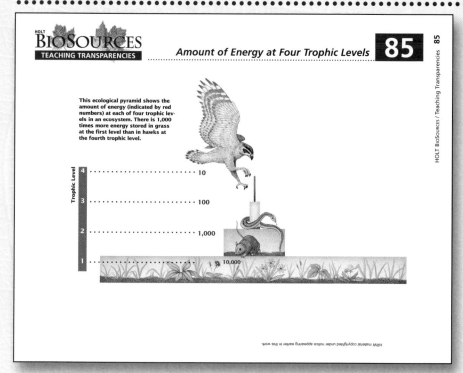

BIOSOURCES
TEACHING TRANSPARENCIES
Amount of Energy at Four Trophic Levels 85

HOLT BioSources / Teaching Transparencies 85

This ecological pyramid shows the amount of energy (indicated by red numbers) at each of four trophic levels in an ecosystem. There is 1,000 times more energy stored in grass at the first level than in hawks at the fourth trophic level.

Trophic Level
4 10
3 100
2 1,000
1 10,000

Teaching Strategies

- Use this transparency when discussing the number of trophic levels in an ecosystem. Review the terms *producer, consumer,* and *decomposer.*
- Explain that this ecological pyramid reflects greater productivity at the producer level than at the consumer levels. Have students calculate the percentage of energy available from one level to the next higher level in the pyramid *(about 10 percent).*
- Explain how the Second Law of Thermodynamics relates to the ecological pyramid and to the energy available from one level to the next.

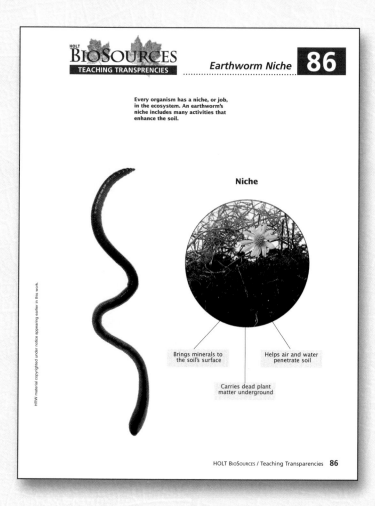

BIOSOURCES
TEACHING TRANSPRENCIES
Earthworm Niche 86

Every organism has a niche, or job, in the ecosystem. An earthworm's niche includes many activities that enhance the soil.

Niche

Brings minerals to the soil's surface

Helps air and water penetrate soil

Carries dead plant matter underground

Earthworm Niche 86

Teaching Strategies

- Use this transparency when discussing the concept of a niche.
- Have students write a list of questions that examine the earthworm's niche. For example, what do earthworms eat, and how do they find food? Where do earthworms live, and what extremes of temperature and sunlight can they withstand? Write the questions on the board and have students discuss possible answers.
- Inform students about earthworm farms and explain why earthworm farms have become popular with farmers and gardeners. Emphasize the vital role earthworms play in composting vegetative matter.

Teaching Strategies

- Use this transparency when discussing how competing organisms evolve. Review the terms *competition, fundamental niche*, and *realized niche*.

- Walk students through Joseph Connell's experiments. Emphasize that competition often prevents an organism from occupying all of its fundamental niche.

- Have students write a paragraph that explains why no two species can occupy the same niche at the same time.

Effect of Competition on an Organism's Niche **87**

a The barnacle *Chthamalus stellatus* can live in both shallow and deep water on a rocky coast. These areas are its fundamental niche.

b The barnacle *Balanus balanoides* prefers to live in deep water, which is its fundamental niche.

c When the two barnacles live together, *Chthamalus* is restricted to shallow water, its realized niche. What is the realized niche of *Balanus*?

Unit 5 | *Viruses, Bacteria, Protists, and Fungi*

88 Two Kingdoms of Prokaryotes

Teaching Strategies

- Use this transparency when discussing the two prokaryotic kingdoms—Archaebacteria and Eubacteria. Review the characteristics of prokaryotic cells, and emphasize that these cells are the oldest life-forms on Earth.

- Point out the extreme conditions that characterize the environments shown in the photos. Explain that archaebacteria live only in very harsh environments, thriving under anaerobic conditions, high salinity, or high temperature and acidity. Add that eubacteria have been evolving for billions of years; their range of habitat is greater than that of any eukaryotes.

- Inform students that thermoacidophilic archaebacteria, such as *Sulfolobus*, require elemental sulfur in their metabolism and that they thrive in hot sulfur springs and volcanic vents.

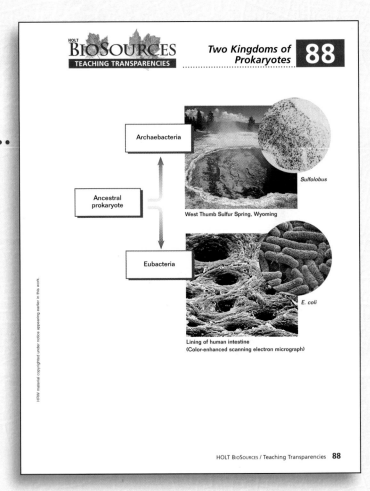

Two Kingdoms of Prokaryotes **88**

Archaebacteria

Sulfolobus

West Thumb Sulfur Spring, Wyoming

Ancestral prokaryote

Eubacteria

E. coli

Lining of human intestine
(Color-enhanced scanning electron micrograph)

HOLT BioSources / Teaching Transparencies **88**

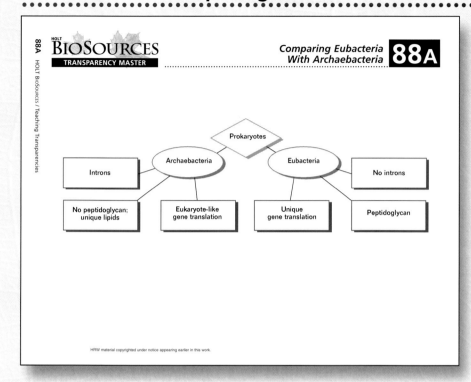

88A

HOLT BioSources / Teaching Transparencies

HOLT
BIOSOURCES
TRANSPARENCY MASTER

Comparing Eubacteria
With Archaebacteria 88A

Prokaryotes

Archaebacteria

Eubacteria

Introns

No introns

No peptidoglycan;
unique lipids

Eukaryote-like
gene translation

Unique
gene translation

Peptidoglycan

HRW material copyrighted under notice appearing earlier in this work.

Teaching Strategies

- Use this transparency when comparing the characteristics of eubacteria with those of archaebacteria. Review the terms *intron, lipid*, and *gene translation*.

- Review the characteristics of each kingdom. Ask students which kingdom has more in common with eukaryotes *(Archaebacteria)*.

- Ask students to estimate how closely related the two prokaryotic kingdoms are. Inform them that molecular sequencing of ribosomal RNA from a variety of organisms suggests that the archaebacteria and the eubacteria are as distantly related to each other as they are to the eukaryotes.

Three Bacterial Cell Shapes 89

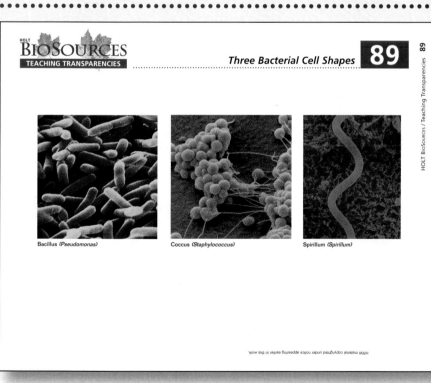

HOLT
BIOSOURCES
TEACHING TRANSPARENCIES

Three Bacterial Cell Shapes 89

89

HOLT BioSources / Teaching Transparencies

Bacillus *(Pseudomonas)*   Coccus *(Staphylococcus)*   Spirillum *(Spirillum)*

HRW material copyrighted under notice appearing earlier in this work.

Teaching Strategies

- Use this transparency when describing bacterial cell shape. Explain that some eubacterial classification systems are based on cell shape.

- Review the three different cell shapes. Have students study the photomicrographs to determine the meaning of the name of each cell shape. *(The bacilli are rod-shaped; the cocci are spherical; and the spirilla are helically coiled.)*

- Ask students to explain why some species of bacteria are found in pairs, chains, or clusters. *(When cell division takes place, the two new cells remain attached and form characteristic groupings.)*

Members of the Kingdom Protista

Teaching Strategies

- Use this transparency when discussing the diversity of the kingdom Protista. Emphasize the characteristics that unite the members of this kingdom. *(All are eukaryotic and lack tissue differentiation.)*

- Briefly review the types of protists listed in the table, and cite the approximate number of species. Have students copy the table onto a sheet of paper and add a column labeled *Characteristics*. As students continue their studies of protists, have them add the characteristics of each group to the table.

- Ask students to name the protists that sound similar, judging on common name alone *(brown algae, green algae, and red algae; cellular slime molds, plasmodial slime molds, and water molds).*

BioSources
TRANSPARENCY MASTER

Members of the
Kingdom Protista 89A

Common Name	Approximate Number of Species
Amoebas	300
Brown algae	1,500
Cellular slime molds	70
Chytrids	575
Ciliates	8,000
Diatoms	more than 11,500
Dinoflagellates	2,100
Euglenoids	1,000
Foraminiferans (Forams)	300
Green algae	more than 7,000
Plasmodial slime molds	500
Red algae	4,000
Sporozoans	3,900
Unicellular flagellates	3,000
Water molds	580

90 Gram Staining

Teaching Strategies

- Use this transparency to illustrate how Gram-staining can be used to identify bacteria. Review the basic steps of the staining procedure.

- Point out the differences in bacterial cell wall structure, emphasizing the difference in thickness of the peptidoglycan layer and the presence or absence of an outer membrane. Ask students to explain why Gram-positive bacteria appear purple. *(These bacteria have a thicker layer of peptidoglycan that resists being broken down by the alcohol wash.)*

- Inform students that Gram-staining is an important diagnostic tool because there are few visible structures that can be used in identifying eubacteria.

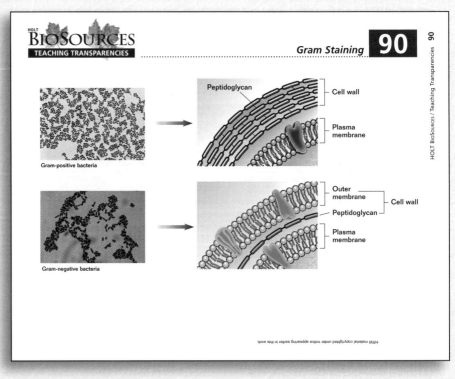

Phylum		
Move using pseudopodia	Rhizopoda (amoebas)	Heterotrophic
	Foraminifera (forams)	
Have double shells made of silica	Bacillariophyta (diatoms)	Photosynthetic
Photosynthetic protists; can be multicellular	Chlorophyta (green algae)	Photosynthetic
	Rhodophyta (red algae)	
	Phaeophyta (brown algae)	
Move using flagella	Dinoflagellata (dinoflagellates)	Photosynthetic
	Zoomastigina (unicellular flagellates)	Heterotrophic
	Euglenophyta (euglenoids)	Most are heterotrophic; some are photosynthetic
Move using cilia	Ciliophora (ciliates)	Heterotrophic
Funguslike protists	Acrasiomycota (cellular slime molds)	Heterotrophic
	Myxomycota (plasmodial slime molds)	
	Oomycota (oomycetes)	
	Chytridiomycota (chytrids)	
Form resistant spores	Sporozoa (sporozoans)	Heterotrophic

Phyla of Protists 90A

Teaching Strategies

• Use this transparency when discussing the phyla that constitute the kingdom Protista. Review the terms *heterotrophic* and *photosynthetic*.

• Describe the distinguishing features of each phyla listed in the table. Inform students that this table is an artificial classification scheme that groups organisms according to their structural similarities. It is not necessarily based on evolutionary relationships.

• Tell students that for practical purposes, protists can be seen as animal-like, plantlike, or funguslike. Have them determine how the listed phyla are divided among these three categories.

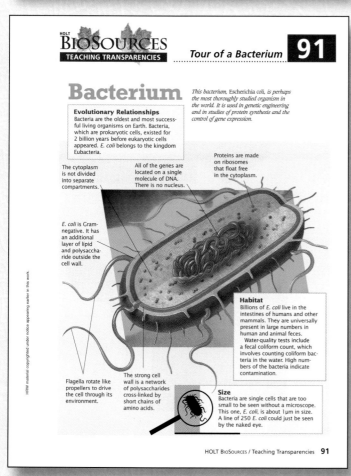

Bacterium

This bacterium, Escherichia coli, *is perhaps the most thoroughly studied organism in the world. It is used in genetic engineering and in studies of protein synthesis and the control of gene expression.*

Evolutionary Relationships
Bacteria are the oldest and most successful living organisms on Earth. Bacteria, which are prokaryotic cells, existed for 2 billion years before eukaryotic cells appeared. *E. coli* belongs to the kingdom Eubacteria.

The cytoplasm is not divided into separate compartments.

All of the genes are located on a single molecule of DNA. There is no nucleus.

Proteins are made on ribosomes that float free in the cytoplasm.

E. coli is Gram-negative. It has an additional layer of lipid and polysaccharide outside the cell wall.

Habitat
Billions of *E. coli* live in the intestines of humans and other mammals. They are universally present in large numbers in human and animal feces. Water-quality tests include a fecal coliform count, which involves counting coliform bacteria in the water. High numbers of the bacteria indicate contamination.

Flagella rotate like propellers to drive the cell through its environment.

The strong cell wall is a network of polysaccharides cross-linked by short chains of amino acids.

Size
Bacteria are single cells that are too small to be seen without a microscope. This one, *E. coli*, is about 1μm in size. A line of 250 *E. coli* could just be seen by the naked eye.

Tour of a Bacterium 91

Teaching Strategies

• Use this transparency to review the characteristics of the eubacterium *Escherichia coli.*

Is *E. coli* Gram-negative or Gram-positive?
(Gram-negative)

Where is this bacterium found, and how does it move through its environment?
(E. coli *lives in the intestines of humans and other mammals. It moves by rotating its flagella like propellers.)*

Why do scientists know so much about this bacterium?
(It is easy to study and has been the subject of many genetic experiments.)

• Inform students that *E. coli* bacteria are used in the Ames test, which involves exposing bacteria to potential carcinogens to see if mutations result. The only drawback to the Ames test is that it relies on the assumption that what is a mutagen for *E. coli* is a carcinogen for humans.

92 Bacterial Diseases and Modes of Transmission

Teaching Strategies

- Use this transparency when discussing important bacterial diseases and their modes of transmission. Inform students that disease-causing bacteria belong to the kingdom Eubacteria.

- Review the bacterial diseases listed in the table, describing their symptoms and mode of transmission. Have students list additional human diseases caused by bacteria. *(Other bacterial diseases and infections might include syphilis, gonorrhea, whooping cough, bacterial pneumonia, bacterial dysentery, meningitis, strep throat, boils, and abscesses.)* Inform students that an equally long list could be compiled of bacterial diseases of other animals and of plants.

- Ask students to describe how they can prevent developing most of these bacterial diseases *(by avoiding infected animals and contaminated water or food).*

Disease	Mode of Transmission	Symptoms
Tuberculosis	Airborne water droplets	Fatigue, persistent cough, bleeding in lungs; can be fatal
Diphtheria	Airborne water droplets	Fever, sore throat, fatigue
Scarlet fever	Airborne water droplets	Rash, fever, sore throat
Bubonic plague	Fleas	Swollen glands, bleeding under skin; often fatal
Typhus	Lice	Rash, chills, fever; often fatal
Tetanus	Dirty wounds	Severe, prolonged muscle spasms
Cholera	Contaminated water	Severe diarrhea, vomiting; often fatal
Typhoid	Contaminated water and food	Headaches, fever, diarrhea, rash; often fatal
Leprosy	Personal contact	Nerve damage, skin lesions, tissue degeneration
Lyme disease	Ticks	Rash, pain, swelling in joints

93 Structures of Adenovirus and Bacteriophage

Teaching Strategies

- Use this transparency when discussing the structure of an adenovirus and of a bacteriophage. Review the term *capsid.*

- Inform students that the capsid of the adenovirus is a polyhedron consisting of 20 equal sides. Adenovirus multiplies in the upper respiratory tract of humans to produce coldlike symptoms.

- Have students compare the adenovirus to the more complex bacteriophage. Explain that the long leglike fibers enable bacteriophage to attach to a host cell *(usually E. coli).*

- Ask students to describe the historical experiment that used bacteriophage to infect bacteria. *(Students should describe the Hershey–Chase experiment, in which radio-labeled bacteriophage were used to identify the genetic material.)*

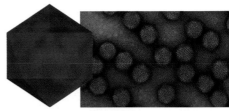

Adenovirus

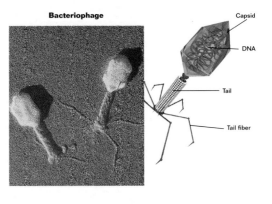

Bacteriophage

Capsid

DNA

Tail

Tail fiber

BioSources
TEACHING TRANSPARENCIES

Structures of TMV and Influenza Virus **94**

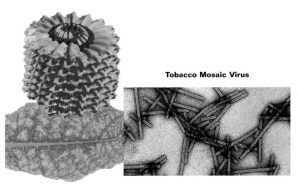

Tobacco Mosaic Virus

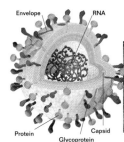

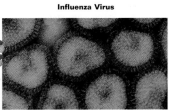

Envelope   RNA

Influenza Virus

Protein   Capsid
        Glycoprotein

Teaching Strategies

- Use this transparency when discussing the structure of a tobacco mosaic virus and of an influenza virus.
- Have students compare the structures of the two viruses in this figure. Point out the lipid envelope of the influenza virus. Inform students that at the center of the cylindrical tobacco mosaic virus is a long RNA molecule.
- Inform students that the tobacco mosaic virus was the first virus to be identified. In 1935 W. M. Stanley of the Rockefeller Institute isolated and crystallized tobacco mosaic virus and showed that if the crystals were injected into tobacco plants, they became active, multiplied, and caused disease symptoms in the plants.

BioSources
TEACHING TRANSPARENCIES

Tour of a Virus **95**

Virus *The disease AIDS is caused by a virus called HIV, human immunodeficiency virus.*

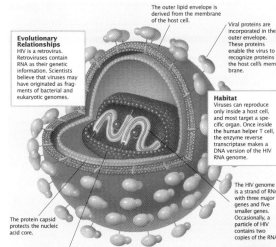

Evolutionary Relationships
HIV is a retrovirus. Retroviruses contain RNA as their genetic information. Scientists believe that viruses may have originated as fragments of bacterial and eukaryotic genomes.

The outer lipid envelope is derived from the membrane of the host cell.

Viral proteins are incorporated in the outer envelope. These proteins enable the virus to recognize proteins in the host cell's membrane.

Habitat
Viruses can reproduce only inside a host cell, and most target a specific organ. Once inside the human helper T cell, the enzyme reverse transcriptase makes a DNA version of the HIV RNA genome.

The HIV genome is a strand of RNA with three major genes and five smaller genes. Occasionally, a particle of HIV contains two copies of the RNA.

The protein capsid protects the nucleic acid core.

The protein core surrounds the RNA, which is the genetic material of HIV.

Size
Viruses are directly comparable to molecules in size. A hydrogen atom is about 0.1 nm. HIV is approximately 100 nm in diameter.

H

Teaching Strategies

- Use this transparency when discussing the characteristics of HIV, the human immunodeficiency virus. Review the terms *retrovirus* and *RNA transcriptase*.
- Discuss the characteristics of HIV.

Where does HIV reproduce?
(usually inside the helper T cell in humans)

Where does the virus get its outer lipid membrane?
(from the membrane of the host cell)

What disease does this virus cause?
(AIDS)

Teaching Strategies

- Use this transparency to discuss the reproductive cycle of HIV. Start with the virus binding to the cell's receptor proteins. Inform students that once inside the cell, HIV sheds its envelope and capsid, leaving only its RNA in the cytoplasm. Review the terms *endocytosis, receptor protein, exocytosis,* and *protein synthesis.*

How is HIV's RNA transcribed into DNA?

(The virus has the enzyme reverse transcriptase, which transcribes viral RNA into a complementary DNA strand.)

What happens to the DNA copy of HIV's genetic information?

(Its genes are translated into HIV proteins.)

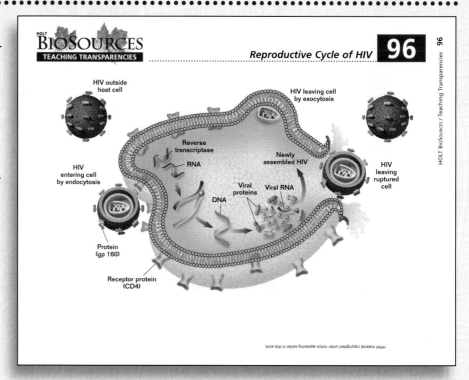

BIOSOURCES
TEACHING TRANSPARENCIES

Reproductive Cycle of HIV **96**

HIV outside host cell

HIV leaving cell by exocytosis

HIV entering cell by endocytosis

Reverse transcriptase

RNA

Newly assembled HIV

HIV leaving ruptured cell

Viral proteins

Viral RNA

DNA

Protein (gp 160)

Receptor protein (CD4)

Teaching Strategies

- Use this transparency to discuss important viral diseases and their modes of transmission.
- Describe the symptoms and mode of transmission of each of the viral diseases listed in the table.

Why don't most viral infections respond to treatment with antibiotics?

(Most antibiotics act by disrupting metabolic pathways or cell membranes. Since viruses have neither of these features, they are not inhibited by antibiotics.)

Why can recovery from a viral infection be so prolonged?

(The immune system can attack the viruses only as they are moving from cell to cell.)

BIOSOURCES
TEACHING TRANSPARENCIES

Viral Diseases and Modes of Transmission **97**

Disease	Transmitted by	Symptoms
Chickenpox	Air currents	Rash, fever
Measles	Air currents	Blotchy rash, high fever, congestion in nose and throat
Rubella (German measles)	Air currents	Rash, swollen glands
Mumps	Air currents	Swollen salivary glands
Influenza (flu)	Air currents	Headache, muscle aches, sore throat, cough; historically, one of the great "killer" diseases
Smallpox	Air currents	High fever, pustules on skin; often fatal; now eliminated
Infectious hepatitis	Contaminated food or water	Fever, chills, nausea, swollen liver, jaundice, pain in the joints
Polio	Contaminated food or water	Headache, stiff neck, possible paralysis
Yellow fever	Mosquitoes	Nausea, fever, aches, liver cell destruction; can be fatal
AIDS	Sexual contact, contaminated blood products, contaminated hypodermic needles and syringes	Immune system failure; fatal

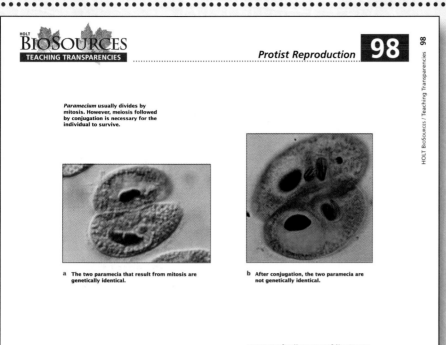

Teaching Strategies

- Use this transparency when describing one form of reproduction found in certain types of protists. Review the terms *mitosis, conjugation,* and *meiosis.*

- Have students describe the process of conjugation. Inform students that the kingdom Protista contains an extremely diverse group of organisms, some reproducing differently.

- Emphasize that paramecia usually reproduce by mitosis, which results in genetically identical cells. Have students explain why it can be disadvantageous for an organism to produce genetically identical offspring. *(It doesn't provide new gene combinations that allow the species to adapt to environmental changes.)*

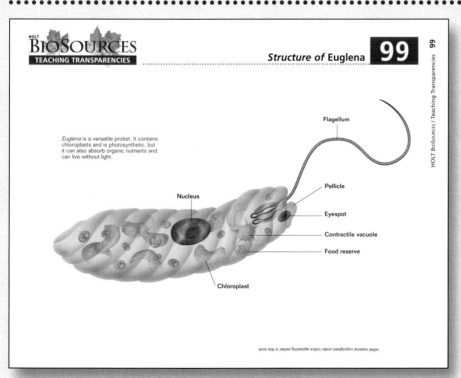

Teaching Strategies

- Use this transparency when discussing the characteristics of the protist Euglena. Review the terms *photosynthetic* and *chloroplast.*

- Describe the characteristics of *Euglena.* Explain that this protist differs from others in that it has flexible internal plates called a pellicle that make up the outer surface.

- Have students design a graphic organizer that delineates *Euglena's* plant-like characteristics and its animal-like characteristics. *(Euglena is plantlike in that it has chloroplasts and is photosynthetic; it is animal-like in that it lacks a cell wall, can be heterotrophic, and is highly motile.)*

Tour of a Protist (Paramecium)

Teaching Strategies

- Use this transparency when describing the characteristics of the protist Paramecium. Review the terms *cilia, flagella, endocytosis,* and *exocytosis.*
- Discuss the characteristics of *Paramecium.*

What kind of protist is *Paramecium*?

(a ciliate)

Where is *Paramecium* found, and how does it move through its environment?

(Paramecium lives in streams and ponds, and it uses its cilia to propel itself through the water.)

How does this protist rid itself of excess water?

(It uses its contractile vacuole to squeeze out excess water.)

- Have students explain why biologists call *Paramecium* organisms without cell boundaries.

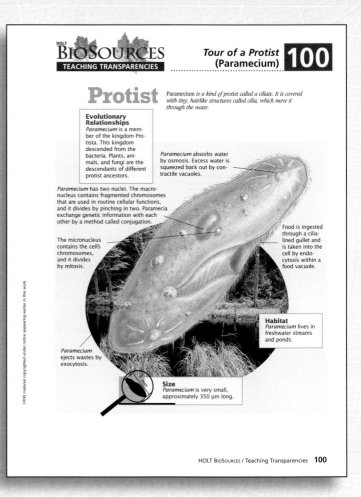

101 Life Cycle of Plasmodium

Teaching Strategies

- Use this transparency when describing the life cycle of the protist *Plasmodium.* Review the terms *zygote, sporozoite, merozoite,* and *gametocyte.*
- Proceed through the life cycle of the sporozoan *Plasmodium.* Emphasize that this parasite spends part of its life cycle inside the female Anopheles mosquito. Add that efforts have been made to eradicate populations of this mosquito species, either directly by the use of insecticides or indirectly by destroying its breeding grounds.
- Describe the symptoms of malaria, a disease caused by *Plasmodium.* Explain that the malaria victim experiences attacks of violent chills and high fevers. Malaria can be an especially serious condition in young children.

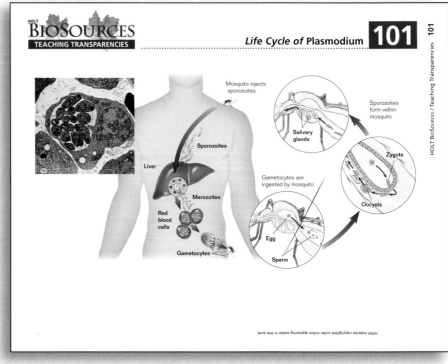

BioSources
TEACHING TRANSPARENCIES

Diseases Caused by Protists **102**

Disease	Host	Organism
Amoebic dysentery	Humans	*Entamoeba*
Malaria	Humans	*Plasmodium*
Toxoplasmosis	Humans, cats	*Toxoplasma*
Giardiasis	Humans	*Giardia*
Sleeping sickness	Humans, tsetse flies	*Trypanosoma*
Leishmaniasis	Humans, sand flies	*Leishmania*
Late blight	Potatoes	*Phytophthora*

Teaching Strategies

• Use this transparency when discussing notable diseases caused by protists and their modes of transmission.

• Review the diseases listed in the table, describing their symptoms and modes of transmission.

Reproduction of Chlamydomonas 103

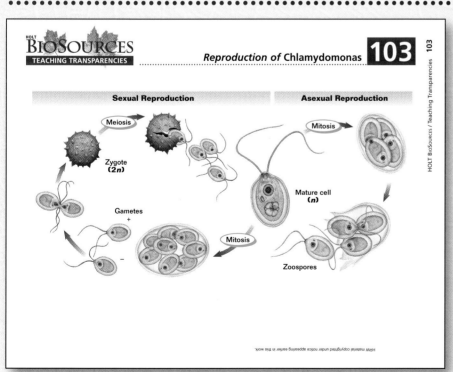

BioSources
TEACHING TRANSPARENCIES

Reproduction of Chlamydomonas **103**

HOLT BioSources / Teaching Transparencies **103**

Sexual Reproduction

Meiosis

Zygote **(2n)**

Gametes +

−

Asexual Reproduction

Mitosis

Mature cell **(n)**

Mitosis

Zoospores

Teaching Strategies

• Use this transparency when discussing the life cycle of the unicellular green alga, *Chlamydomonas*. Review the terms *sexual reproduction, asexual reproduction, meiosis, mitosis,* and *zygote.* Explain the concept of *mating type.*

• Proceed through the life cycle, emphasizing that *Chlamydomonas* is able to reproduce sexually and asexually. Have students describe the step that differentiates sexual reproduction from asexual reproduction. *(In sexual reproduction, haploid gametes of opposite mating types fuse to form a diploid zygote, which undergoes meiosis.)*

Teaching Strategies

- Use this transparency when discussing the life cycle of the multicellular green alga *Ulva*, commonly called sea lettuce. Review the terms *sporophyte* and *gametophyte*.

- Proceed through the events in the life cycle, emphasizing that Ulva's life cycle is sexual. Have students identify where fusion and meiosis take place.

- Explain that the haploid and diploid forms of *Ulva* look the same but differ in their chromosome number. Ask students to explain the term *alternation of generations*, and have them design a graphic organizer that defines the two forms of *Ulva*.

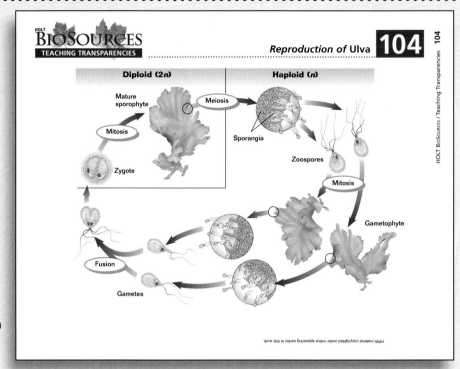

HOLT BIOSOURCES TEACHING TRANSPARENCIES

Reproduction of Ulva **104**

HOLT BioSources / Teaching Transparencies **104**

Teaching Strategies

- Use this transparency when discussing the following six divisions of algae: Chlorophyta, Phaeophyta, Rhodophyta, Bacillariophyta, Dinoflagellata, and Euglenophyta.

- Review the characteristics of the six divisions of algae, pointing out that all of the organisms listed in this table are autotrophic. Have students explain how these organisms obtain food. *(All contain photosynthetic pigments, including chlorophyll a.)*

HOLT BIOSOURCES TEACHING TRANSPARENCIES

Six Divisions of Algae **105**

HOLT BioSources / Teaching Transparencies **105**

Division	Cell Types	Photosynthetic Pigments	Form of Food Storage	Cell Wall Composition
Chlorophyta (*Chloro* = "green"): 7,000 species	Both unicellular and multicellular	Chlorophylls a and b, xanthophylls, carotenes	Starch	Polysaccharides, sometimes cellulose
Phaeophyta (*Phaeos* = "brown"): 1,500 species	Mostly multicellular	Chlorophylls a and c, fucoxanthin (a carotene)	Laminarin (oil)	Cellulose with alginic acids
Rhodophyta (*Rhodo* = "red"): 4,000 species	Mostly multicellular	Chlorophylls a and d, phycobilins	Starch	Cellulose or pectin, many with calcium carbonate
Bacillariophyta (*Bacillus* = "rod"): 10,000 species	Mostly unicellular	Chlorophylls a and c, carotenes, including fucoxanthin	Chrysolaminarin (oily carbohydrate)	Cellulose, some with silica; possibly no cell wall
Dinoflagellata (*Dino* = "whirling"): 1,100 species	All unicellular	Chlorophylls a and c, carotenes	Starch	Cellulose
Euglenophyta (*Euglena* = "true eye"):800 species	Mostly unicellular	Chlorophylls a and b, carotenes in genera with chloroplasts	Paramylon (a starch)	No cell wall; protein-rich pellicle

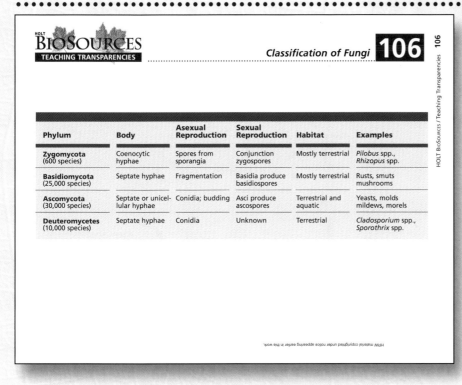

Phylum	Body	Asexual Reproduction	Sexual Reproduction	Habitat	Examples
Zygomycota (600 species)	Coenocytic hyphae	Spores from sporangia	Conjunction zygospores	Mostly terrestrial	*Pilobus* spp., *Rhizopus* spp.
Basidiomycota (25,000 species)	Septate hyphae	Fragmentation	Basidia produce basidiospores	Mostly terrestrial	Rusts, smuts mushrooms
Ascomycota (30,000 species)	Septate or unicellular hyphae	Conidia; budding	Asci produce ascospores	Terrestrial and aquatic	Yeasts, molds mildews, morels
Deuteromycetes (10,000 species)	Septate hyphae	Conidia	Unknown	Terrestrial	*Cladosporium* spp., *Sporothrix* spp.

HRW material copyrighted under notice appearing earlier in this work.

Teaching Strategies

- Use this transparency when describing the four divisions of fungi. Review the terms *hypha, septum, spores,* and *conjugation.*
- Review the characteristics of the four division of fungi, emphasizing that fungi are classified according to their sexual reproductive structures. Ask students to name the classification division for species that have no observable form of sexual reproduction *(Deuteromycota).*
- Have students determine what distinguishes the hyphae of a zygomycete from those of a basidiomycete. *(Zygomycete have coenocytic hyphae; basidiomycetes have septate hyphae.)*

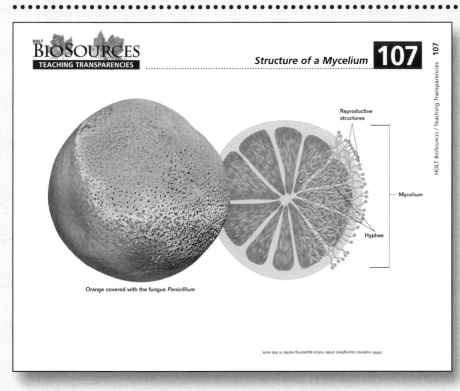

Reproductive structures

Mycelium

Hyphae

Orange covered with the fungus *Penicillium*

HRW material copyrighted under notice appearing earlier in this work.

Teaching Strategies

- Use this transparency when discussing the structure of a fungal mycelium. Review the terms *hypha* and *mycelium.*
- Explain that the body of a fungus is composed of threadlike hyphae, which branch throughout the food supply. The visible part of the fungus is usually the reproductive structures; an example is the green-and-white fuzz covering the orange in the illustration.
- Have students explain why the structure of a fungus is well suited to its function. *(A mycelium creates a high surface-area-to-volume ratio, which is essential for absorbing food from the environment.)*

Teaching Strategies

- Use this transparency to discuss the life cycle of *Rhizopus*, a black bread mold. Review the terms *hypha, zygosporangium, sporangia*, and *spore*.

- Highlight the points at which fusion, meiosis, and spore release take place. Emphasize that sexual reproduction occurs when hyphae from opposite mating types grow together and fuse. Ask students to explain what happens to pairs of haploid nuclei after two hyphae fuse. *(The haploid nuclei fuse to produce diploid nuclei, or zygotes.)*

- Inform students that zygomycetes usually reproduce asexually and that thousands of asexual spores can be carried long distances by wind, rain, and animals.

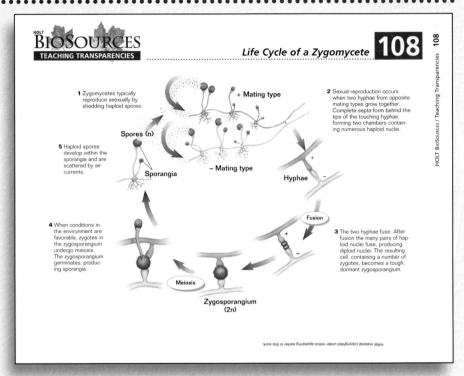

Teaching Strategies

- Use this transparency when discussing the life cycle of a typical ascomycete. Review the terms *ascus* and *ascocarp*.

- Review each step in the life cycle, highlighting the points at which meiosis, mitosis, spore release, and ascus formation take place. Point out that unlike zygomycetes, ascomycetes are composed of hyphae that have cross walls called septa.

- Inform students that ascomycetes are very diverse, varying from unicellular yeasts, powdery mildews, and cottony molds to morel, truffles, and cup fungi. Explain that although yeast, which are unicellular, usually reproduce by budding, they will form asci when exposed to unfavorable environmental conditions.

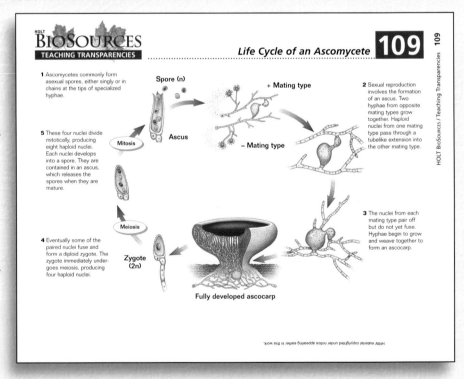

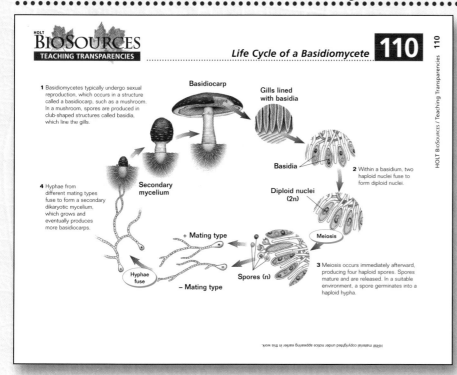

1 Basidiomycetes typically undergo sexual reproduction, which occurs in a structure called a basidiocarp, such as a mushroom. In a mushroom, spores are produced in club-shaped structures called basidia, which line the gills.

Basidiocarp

Gills lined with basidia

Basidia

2 Within a basidium, two haploid nuclei fuse to form diploid nuclei.

Secondary mycelium

Diploid nuclei (2n)

Meiosis

4 Hyphae from different mating types fuse to form a secondary dikaryotic mycelium, which grows and eventually produces more basidiocarps.

+ Mating type

Hyphae fuse

Spores (n)

3 Meiosis occurs immediately afterward, producing four haploid spores. Spores mature and are released. In a suitable environment, a spore germinates into a haploid hypha.

– Mating type

Teaching Strategies

- Use this transparency when discussing the life cycle of a typical basidiomycete, such as a mushroom. Review the terms *basidiocarp, basidium*, and *dikaryotic*.

- Review each step in the life cycle, highlighting the point at which meiosis, spore release, fusion, and basidiocarp formation take place. Have students explain the basidiomycete life cycle in their own words.

- Inform students that mushrooms are just the "tip of the iceberg." Like all other members of the kingdom Fungi, the main part of the fungus is the mycelium, a woven mat of hyphae hidden below the soil surface.

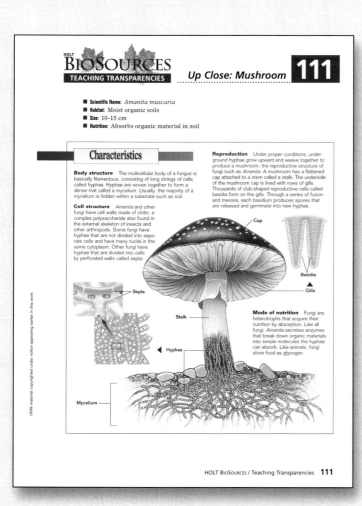

- **Scientific Name:** *Amanita muscaria*
- **Habitat:** Moist organic soils
- **Size:** 10–15 cm
- **Nutrition:** Absorbs organic material in soil

Characteristics

Body structure The multicellular body of a fungus is basically filamentous, consisting of long strings of cells called hyphae. Hyphae are woven together to form a dense mat called a mycelium. Usually, the majority of a mycelium is hidden within a substrate such as soil.

Cell structure *Amanita* and other fungi have cell walls made of chitin, a complex polysaccharide also found in the external skeleton of insects and other arthropods. Some fungi have hyphae that are not divided into separate cells and have many nuclei in the same cytoplasm. Other fungi have hyphae that are divided into cells by perforated walls called septa.

Reproduction Under proper conditions, underground hyphae grow upward and weave together to produce a mushroom, the reproductive structure of fungi such as *Amanita*. A mushroom has a flattened cap attached to a stem called a stalk. The underside of the mushroom cap is lined with rows of gills. Thousands of club-shaped reproductive cells called basidia form on the gills. Through a series of fusion and meiosis, each basidium produces spores that are released and germinate into new hyphae.

Mode of nutrition Fungi are heterotrophs that acquire their nutrition by absorption. Like all fungi, *Amanita* secretes enzymes that break down organic materials into simple molecules the hyphae can absorb. Like animals, fungi store food as glycogen.

Cap

Basidia

Gills

Septa

Stalk

Hyphae

Mycelium

Up Close: Mushroom 111

Teaching Strategies

- Use this transparency when describing the characteristics of a mushroom *Amanita muscaria*, a representative of the kingdom Fungi. Review the terms *hypha, septum, mycelium, chitin, basidium*, and *heterotroph*.

- Review the basic characteristics of the mushroom, emphasizing that it represents the kingdom Fungi. Emphasize that the mushroom is only a small part of the total organism; the majority of the fungus is a woven mat beneath the soil surface.

- Have students explain why fungi are call external heterotrophs. *(Fungi secrete digestive enzymes that break down organic molecules outside their bodies.)*

Teaching Strategies

- Use this transparency when discussing essential amino acids and complete proteins. Have students identify the source of each food in this meal *(cereal, legume, and animal)*.

- Explain that unlike foods from animals, foods from plants have small amounts of some essential amino acids and large amounts of others. Thus, foods from plants must be eaten in combinations to provide a balance of essential amino acids.

- Have students identify the pairs of foods from plants that provide a complete protein in this meal *(beans and rice, beans and corn tortillas)*.

Which food in this meal provides a complete protein?

(tamales)

HOLT
BIOSOURCES
TEACHING TRANSPARENCIES

Meal Supplying a Complete Protein **112**

Beans (legume that contains some essential amino acids)

Corn tortillas (made of grain; supply some essential amino acids)

Tamales (made of corn and meat; supply essential amino acids)

Rice (grain rich in carbohydrates)

112A *Medicines Originally Derived From Plants*

Teaching Strategies

- Use this transparency when discussing the role of plants in healing and in medicine, both historically and currently. Point out that 80 percent of the people in the world today still use herbs and medicinal plants for primary health care.

- Guide students through the range of medicines from plants and the uses of each. Point out that different plants, as shown in the table, affect different body systems, and note that medicines are made from various parts of plants.

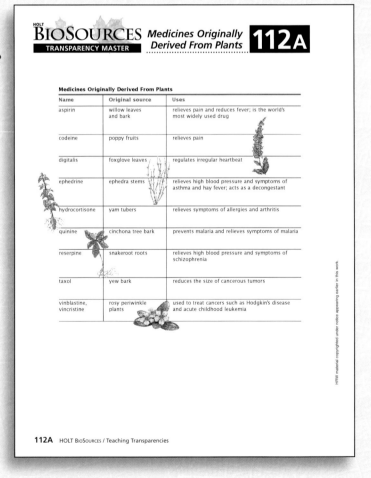

HOLT
BIOSOURCES
TRANSPARENCY MASTER

Medicines Originally Derived From Plants **112A**

Medicines Originally Derived From Plants

Name	Original source	Uses
aspirin	willow leaves and bark	relieves pain and reduces fever; is the world's most widely used drug
codeine	poppy fruits	relieves pain
digitalis	foxglove leaves	regulates irregular heartbeat
ephedrine	ephedra stems	relieves high blood pressure and symptoms of asthma and hay fever; acts as a decongestant
hydrocortisone	yam tubers	relieves symptoms of allergies and arthritis
quinine	cinchona tree bark	prevents malaria and relieves symptoms of malaria
reserpine	snakeroot roots	relieves high blood pressure and symptoms of schizophrenia
taxol	yew bark	reduces the size of cancerous tumors
vinblastine, vincristine	rosy periwinkle plants	used to treat cancers such as Hodgkin's disease and acute childhood leukemia

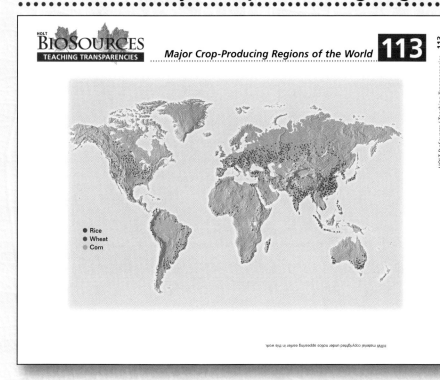

BIOSOURCES
TEACHING TRANSPARENCIES
Major Crop-Producing Regions of the World **113**

HOLT BioSources / Teaching Transparencies 113

- Rice
- Wheat
- Corn

HRW material copyrighted under notice appearing earlier in this work.

Teaching Strategies

- Use this transparency when discussing the major crop-producing regions of the world. Have students locate the regions on the map where rice, wheat, and corn originated.
- Ask students to explain how the distribution of these crops has changed since their origin. *(All are now grown worldwide.)*

Which countries are important producers of rice?

(China, India, and Brazil)
of wheat?

(the United States, Canada, Russia, Ukraine, Poland, Germany, and China)

Which crop is grown in the most countries?

(wheat)

Phyla of Living Plants 113A

BIOSOURCES
TRANSPARENCY MASTER
Phyla of Living Plants **113A**

Phylum	Number of Species	Main Characteristics
Nonvascular Plants		
Hepatophyta Liverworts	6,000	Simplest plants; small, having a dominant gametophyte with a flattened or "leafy" body that lacks vascular tissue, a cuticle, stomata, roots, stems, and leaves
Anthocerophyta Hornworts	100	Small, with a flattened, dominant gametophyte that has stomata but lacks vascular tissue, roots, stems, and leaves
Bryophyta Mosses	10,000	Small; most have simple vascular tissue, a sporophyte consisting of a bare stalk and a spore capsule, and a dominant, "leafy" green gametophyte that lacks roots, stems, and leaves
Vascular Plants		
Psilotophyta Whisk ferns	21	Seedless, with a small, independent gametophyte and a dominant sporophyte that is highly branched and has tiny leaves but is not differentiated into roots and stems
Sphenophyta Horsetails	15	Seedless, with a small, independent gametophyte and a dominant sporophyte consisting of roots and ribbed and jointed stems with soft needlelike leaves at the joints
Lycophyta Club mosses	1,000	Seedless, with a small, independent gametophyte and a dominant, mosslike sporophyte with roots, stems, and leaves

HRW material copyrighted under notice appearing earlier in this work.

113A HOLT BioSources / Teaching Transparencies

Teaching Strategies

- Use this transparency when describing the following six plant phyla: Hepatophyta, Anthocerophyta, Bryophyta, Psilotophyta, Spenophyta, and Lycophyta.
- Have students compare the illustration of the representative of each plant group with the group's characteristics.
- Have students use the table to answer the following questions:

Why must nonvascular plants stay small?

(They lack a vascular system that transports water, minerals, and products of photosynthesis and that provides support.)

What other characteristics do nonvascular plants share?

(All have a dominant gametophyte and lack true roots, stems, and leaves.)

Teaching Strategies

- Use this transparency when discussing the origins of some of the first cultivated crops. Be sure students can locate the agricultural region known as the Fertile Crescent on a world map.
- Discuss the geography of the Fertile Crescent.
- Have students write a paragraph that describes the origins of the early cultivated forms of wheat, barley, and peas.

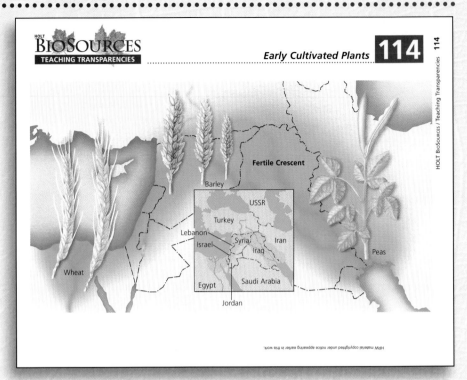

BioSources
TEACHING TRANSPARENCIES

Early Cultivated Plants **114**

Fertile Crescent

Barley

USSR

Turkey

Lebanon

Israel   Syria   Iran

Iraq

Peas

Wheat

Egypt   Saudi Arabia

Jordan

114A
Phyla of Living Plants, continued

Teaching Strategies

- Use this transparency when describing the following six plant phyla: Pterophyta, Coniferophyta, Cycadophyta, Ginkgophyta, Gnetophyta, and Anthophyta.
- Have students compare the illustration of the representative of each plant group with the group's characteristics.
- Have students use the table to answer the following questions:

Which plants need abundant moisture to reproduce? Why?

(plants lacking seeds, because they need abundant moisture to enable sperm to swim to eggs)

Which plants are able to grow tall? Why?

(plants with a vascular system, because they have the ability to distribute water and nutrients over long distances)

BioSources
TRANSPARENCY MASTER

Phyla of Living Plants, continued **114A**

Phylum	Number of Species	Main Characteristics
Pterophyta Ferns	12,000	Seedless, with a small, independent gametophyte and a dominant sporophyte consisting of roots, horizontal stems, and leaves called fronds; spores are produced in clusters of sporangia on lower surfaces of leaves
Coniferophyta Conifers	550	Gymnosperms (seed plants with tiny gametophytes, a large sporophyte, and ovules not enclosed by an ovary) that produce cones; mostly evergreen trees and shrubs with leaves modified as needles or scales
Cycadophyta Cycads	100	Gymnosperms with palmlike leaves; produce male and female cones on separate plants
Ginkgophyta Ginkgo	1	Gymnosperm; deciduous tree with fanlike leaves; produces conelike male reproductive structures and uncovered seeds on separate individuals
Gnetophyta Gnetophytes	70	Gymnosperms; diverse group of shrubs and vines
Anthophyta Flowering plants	250,000	Angiosperms (seed plants with tiny gametophytes, a large sporophyte, and ovules enclosed by an ovary); a very diverse group including trees, shrubs, vines, and herbs that produce flowers and fruits

114A HOLT BioSources / Teaching Transparencies

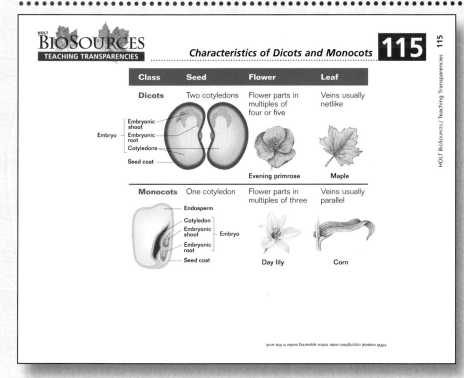

Teaching Strategies

- Use this transparency when summarizing the characteristics of dicots and monocots. Review the term *cotyledon*.

- Inform students that they may have to consider two or more of these characteristics to classify a specific plant as a monocot or a dicot.

- Show students examples or photographs of plants that have easily identifiable characteristics. Include some monocots that have netlike veins, such as pothos ivy, banana, and greenbriar. Have students identify each plant as a monocot or a dicot.

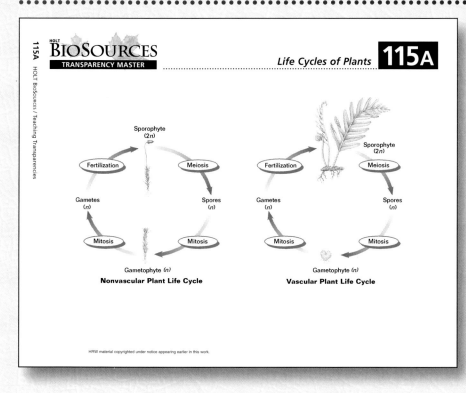

Teaching Strategies

- Use this transparency when discussing the similarities and differences in plant life cycles. Review the terms *sporophyte, gametophyte*, and *alternation of generations*.

- Have students compare the relative sizes of sporophytes and gametophytes in the moss and the fern.

- Ask students to state the number of chromosomes found in a gametophyte. *(n)*

What is the name of this type of life cycle?

(alternation of generations)

How do these life cycles differ?

(The moss gametophyte and sporophyte are about the same size, while the fern's gametophyte is considerably smaller than its sporophyte.)

116 Life Cycle of an Angiosperm

Teaching Strategies

- Use this transparency when discussing the life cycle of angiosperms. Review the terms *gametophyte, pollination, double fertilization,* and *sporophyte.*
- Proceed through the angiosperm life cycle, starting with the mature sporophyte (plant with flowers). Point out the changes that occur in the chromosome number. Also point out that microspores and megaspores form within the male and female parts of the flower, respectively. These spores become microgametophytes (pollen grains) and megagametophytes (containing eggs).
- Have students explain how a sperm reaches the eggs in a flower *(via pollen tubes that grow through the style and into the ovules).*

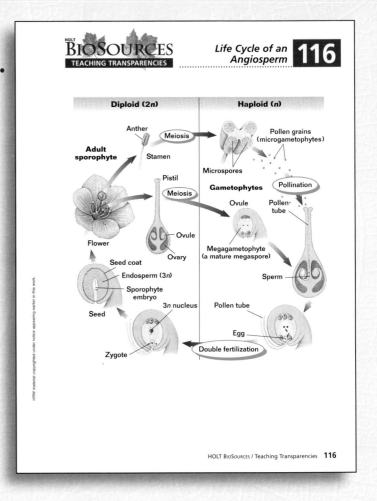

116A Modified Leaves

Teaching Strategies

- Use this transparency when describing different types of modified leaves.
- Have students use the information in the table to answer the following questions:

How do cactus spines help a cactus conserve water?

(They greatly reduce the surface area from which water can evaporate.)

How do tendrils help a garden pea climb?

(They wrap around objects and keep the plant from falling over as it grows taller.)

How do the leaves of a Venus' flytrap help it photosynthesize?

(They are green and contain chlorophyll.)

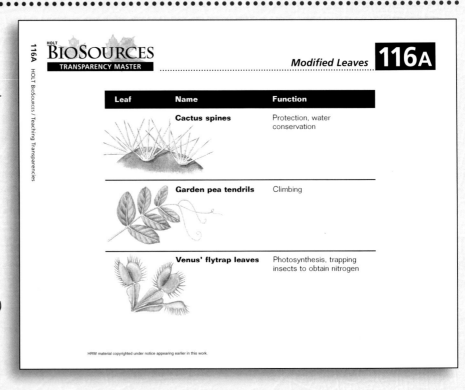

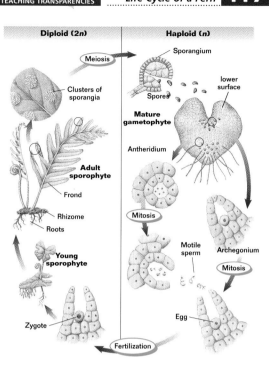

Diploid (2n) | **Haploid (n)**

Meiosis
Sporangium
Spores
lower surface
Clusters of sporangia
Mature gametophyte
Antheridium
Adult sporophyte
Frond
Rhizome
Roots
Mitosis
Motile sperm
Archegonium
Mitosis
Young sporophyte
Egg
Zygote
Fertilization

HOLT BIOSOURCES / Teaching Transparencies **117**

Teaching Strategies

- Use this transparency when discussing the life cycle of ferns. Review the terms *antheridium* and *archegonium*.
- Proceed through the fern life cycle, starting with the mature sporophyte. Point out that when meiosis occurs, the chromosome number changes from diploid to haploid. Also point out that a spore grows into a gametophyte, which is haploid. Finally, note that when fertilization occurs, the chromosome number becomes diploid and a new sporophyte generation begins.
- Ask students to describe where they would look to find a fern gametophyte (*on moist soil beneath a patch of fern sporophytes*).

Modified Stems 117A

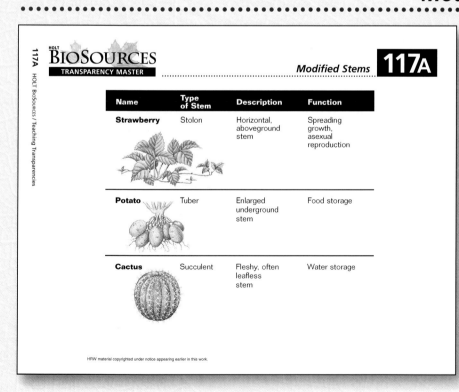

Name	Type of Stem	Description	Function
Strawberry	Stolon	Horizontal, aboveground stem	Spreading growth, asexual reproduction
Potato	Tuber	Enlarged underground stem	Food storage
Cactus	Succulent	Fleshy, often leafless stem	Water storage

Teaching Strategies

- Use this transparency when describing different types of modified stems.
- Have students use the information in the table to answer the following questions:

How do stolons help a strawberry plant spread along the ground?
(*By growing horizontally, stolons increase the diameter of a strawberry plant.*)

How do tubers help a potato plant store food?
(*By holding excess starch and becoming enlarged, tubers are able to store large amounts of food.*)

How does a succulent stem help a cactus store water?
(*Fleshy tissue holds water.*)

118 *Life Cycle of a Moss*

Teaching Strategies

- Use this transparency when discussing the life cycle of mosses. Review the terms *sporangium, antheridium,* and *archegonium*.
- Proceed through the moss life cycle, starting with the mature sporophyte. Point out that when meiosis occurs, the chromosome number changes from diploid to haploid. Also point out that a spore grows into a gametophyte, which is haploid. Finally, point out that when fertilization occurs, the chromosome number becomes diploid and a new sporophyte generation begins.
- Ask students to describe where they would find a moss sporophyte *(at the top of a moss gametophyte).*

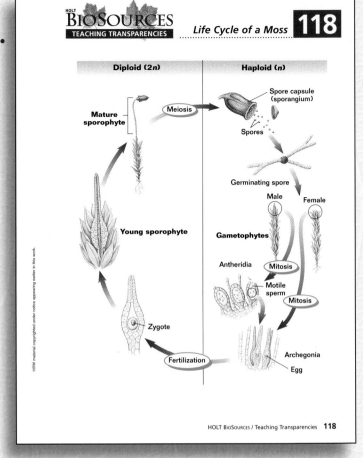

HOLT BIOSOURCES
TEACHING TRANSPARENCIES
Life Cycle of a Moss 118

118A *Types of Plants*

Teaching Strategies

- Use this transparency when discussing the four different plant types—annuals, biennials, herbaceous perennials, and woody perennials. Challenge students to think of other familiar examples of annuals, biennials, and perennials.
- Ask students to explain how annuals differ from biennials. *(Annuals grow, flower, produce seeds, and die within one year or growing season. Biennials take two years or growing seasons to grow, flower, and produce seeds before they die.)*

What are two basic types of woody perennials?

(deciduous and evergreen)

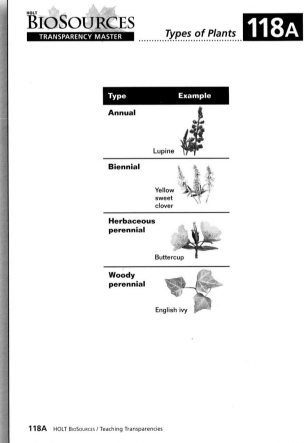

HOLT BIOSOURCES
TRANSPARENCY MASTER
Types of Plants 118A

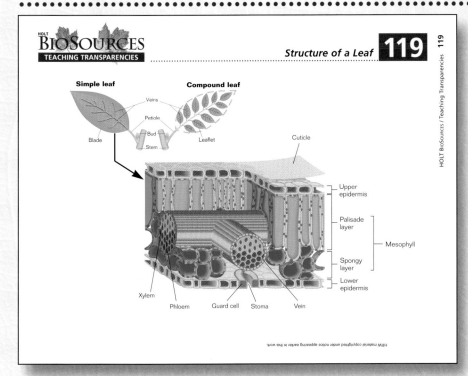

Teaching Strategies

- Use this transparency when discussing the structure of a leaf. Review the terms *stoma, guard cell, mesophyll, xylem,* and *phloem.*

- Have students describe the differences between simple and compound leaves. Point out that one way to tell a leaf from a leaflet is to look for the bud in the axil of the leaf (the angle between a leaf and the stalk or stem to which it is attached).

- Emphasize that there are many variations in leaf structure. For example, plants living in arid climates tend to have thickened cuticles and several layers of epidermal cells. Plants living in areas with intense light tend to have leaves with two or more palisade layers.

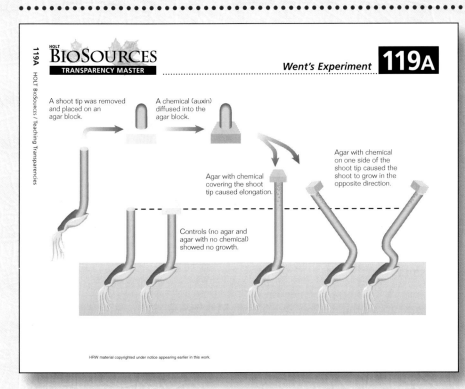

Teaching Strategies

- Use this transparency when describing Frits Went's experiment with oat seedlings. Review the term *agar.*

- Proceed through the steps of Went's experiment. Have students describe how auxin affected the seedling when it entered both sides of the tip. *(The seedling grew but did not bend.)*

- Have students describe how auxin affected the seedling when it entered only one side of the tip. *(The seedling grew and bent toward the opposite side.)*

120 *Plant Cell Structure*

Teaching Strategies

- Use this transparency to point out the organelles and other plant cell structures. Emphasize that most plant cells lack one or more of these features.

- Have students compare plant cell structure with animal cell structure. Ask them to identify which organelles and structures are found in both kinds of cells.

Which organelles and structures are found only in plant cells?

(cell wall and chloroplasts)

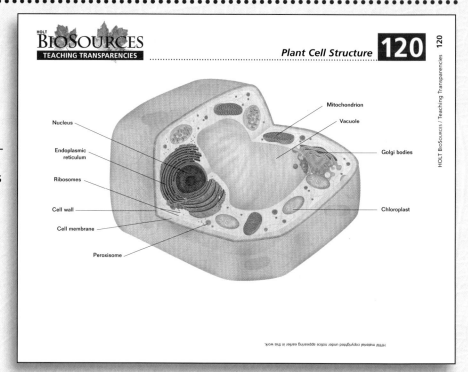

BioSources TEACHING TRANSPARENCIES

Plant Cell Structure **120**

Nucleus
Endoplasmic reticulum
Ribosomes
Cell wall
Cell membrane
Peroxisome
Mitochondrion
Vacuole
Golgi bodies
Chloroplast

120A *Stages of Plant Differentiation*

Teaching Strategies

- Use this transparency when summarizing the stages of plant growth and differentiation. Tell students that each arrow should be read "give (or gives) rise to."

- Ask students to state the type of tissue from which the vascular and cork cambia arise *(differentiated primary tissue)*.

What type of tissue produces primary growth?
(apical meristem)

What type of tissue produces secondary growth?
(vascular and cork cambia)

What is another name for the vascular and cork cambia?
(lateral meristems)

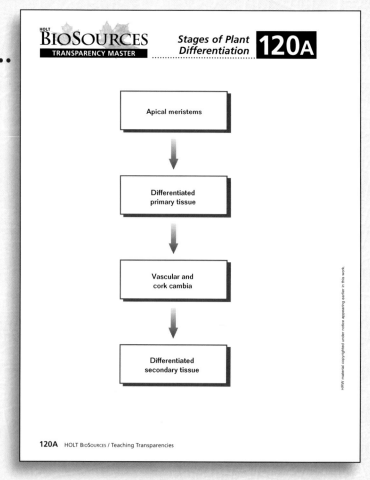

BioSources TRANSPARENCY MASTER

Stages of Plant Differentiation **120A**

Apical meristems

Differentiated primary tissue

Vascular and cork cambia

Differentiated secondary tissue

120A HOLT BioSources / Teaching Transparencies

Structure of a
Vascular Plant **121**

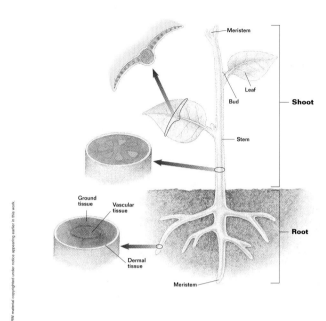

Meristem

Leaf

Bud

Shoot

Stem

Ground
tissue

Vascular
tissue

Root

Dermal
tissue

Meristem

HRW material copyrighted under notice appearing earlier in this work.

HOLT BioSources / Teaching Transparencies **121**

Teaching Strategies

- Use this transparency when reviewing the basic body plan of a vascular plant. Emphasize that most vascular plants have roots and a shoot with stems and leaves. Review the terms *meristem* and *vascular tissue*.
- Point out that the three tissue systems of a vascular plant extend throughout the plant's body.
- Have students locate the plant's dermal tissue. *(Dermal tissue covers the plant's entire body.)*

Where is vascular tissue found in roots?
(in a central cylinder)

Where is ground tissue found in leaves?
(beneath the dermal tissue and surrounding bundles of vascular tissue)

A Closer Look
at a Leaf **122**

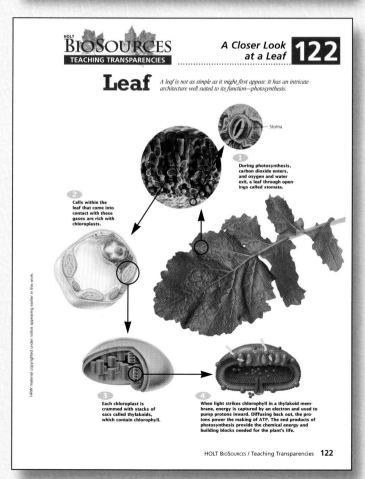

Leaf

A leaf is not as simple as it might first appear. It has an intricate architecture well suited to its function—photosynthesis.

Stoma

1. During photosynthesis, carbon dioxide enters, and oxygen and water exit, a leaf through openings called stomata.

2. Cells within the leaf that come into contact with these gases are rich with chloroplasts.

3. Each chloroplast is crammed with stacks of sacs called thylakoids, which contain chlorophyll.

4. When light strikes chlorophyll in a thylakoid membrane, energy is captured by an electron and used to pump protons inward. Diffusing back out, the protons power the making of ATP. The end products of photosynthesis provide the chemical energy and building blocks needed for the plant's life.

HRW material copyrighted under notice appearing earlier in this work.

HOLT BioSources / Teaching Transparencies **122**

Teaching Strategies

- Use this transparency when discussing the structure and function of a leaf. Review the terms *photosynthesis, chloroplast, chlorophyll*, and *proton pump*.
- Review the structures of the leaf with students, beginning with the stomata. Tell students that oxygen, carbon dioxide, and water enter and exit through these small openings.
- Have students identify the organelle that is the site of photosynthesis *(chloroplast)*.

Where is chlorophyll contained within chloroplasts?
(inside sacs called thylakoids)

- Have students explain in their own words how thylakoid membranes function in manufacturing ATP.

Teaching Strategies

- Use this transparency when describing the structure of stems. Review the terms *vascular bundle, phloem,* and *xylem.*
- Have students compare the external and internal structures of herbaceous and woody stems.
- Ask students what structures occur on both types of stem (*buds, nodes, and internodes*).

What are nodes?
(places where buds and leaves attach to stems)

What structures occur on woody stems but not in herbaceous stems?
(cork and bark)

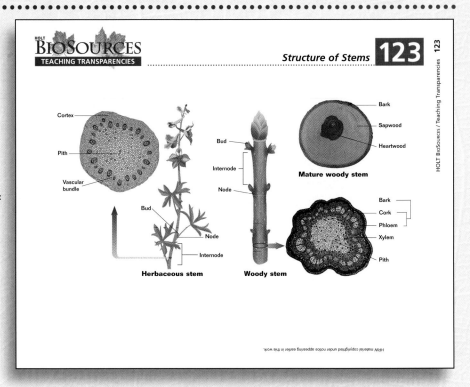

Teaching Strategies

- Use this transparency when describing the structure of roots. Review the terms *primary root, lateral root, root hairs,* and *root cap.*
- Proceed through this figure, emphasizing that the internal structure of a root differs from that of a stem. Tell students that the central cylinder of vascular tissue in a root is called the steele.
- Have students describe the function of a root hair (*to increase surface area for water absorption*).

How can you distinguish a cross section of a root from that of a stem?
(In stems, the vascular tissue is arranged in many bundles. In roots, all of the vascular tissue is found in the center.)

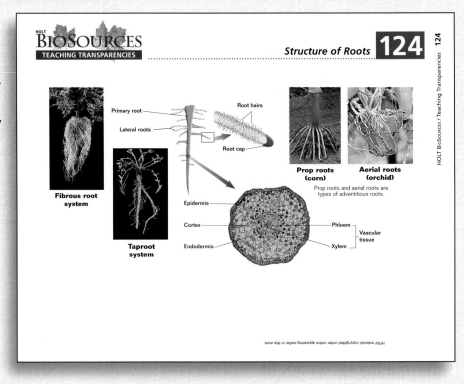

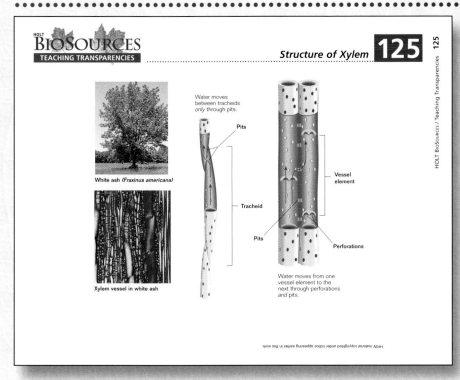

Teaching Strategies

- Use this transparency when discussing the structure of xylem. Review the terms *tracheid* and *vessel element*.
- Have students compare the structure of a tracheid with that of a vessel element. Point out that water can move from one tracheid to the next only through the pits located in the tapered ends of the cells. Emphasize that water can move between adjacent vessel elements through pits in the sides of cells.
- Compare the structure of a xylem vessel to a pipeline consisting of several sections.

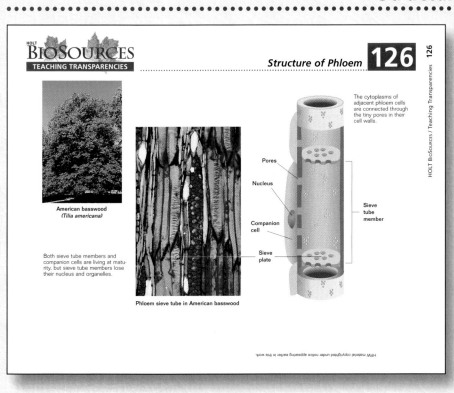

Teaching Strategies

- Use this transparency when discussing the structure of phloem. Review the terms *sieve tube member, companion cell*, and *sieve plate*. Tell students that sieve tubes were named for the clusters of pores, which resemble sieves, in their cells.
- Have students compare phloem sieve tubes with xylem vessels. Point out that both resemble pipelines with several sections of pipe.
- Ask students to describe the major difference between xylem and phloem. *(Mature phloem cells are alive, while mature xylem cells are dead.)*

127 *Pressure Flow Model of Translocation*

Teaching Strategies

- Use this transparency when describing the pressure flow model of translocation. Review the terms *active transport* and *osmosis*.
- Proceed through the diagram. Have students describe in their own words what happens during translocation.
- Have students explain when a root cell can be a source *(when upper parts of the plant need energy and cannot get sugars from the leaves, such as in the spring when new growth begins).*

Are new leaves developing in buds a source or a sink? *(a sink)*

Why? *(New leaves cannot photosynthesize while in a bud and therefore need sugars from elsewhere to make ATP for growth.)*

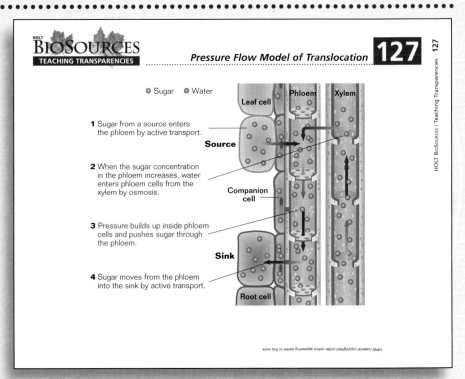

HOLT BIOSOURCES TEACHING TRANSPARENCIES

Pressure Flow Model of Translocation **127**

● Sugar ● Water

1 Sugar from a source enters the phloem by active transport.

Source

2 When the sugar concentration in the phloem increases, water enters phloem cells from the xylem by osmosis.

3 Pressure builds up inside phloem cells and pushes sugar through the phloem.

Sink

4 Sugar moves from the phloem into the sink by active transport.

Leaf cell · Phloem · Xylem · Companion cell · Root cell

---

128 *Events in Germination*

Teaching Strategies

- Use this transparency when describing the events in germination. Review the terms *cotyledon, monocot,* and *dicot.*
- Have students compare the structures of germinating corn and bean seeds. Tell them that the cotyledons of some dicot seeds remain underground.
- Ask students to describe how monocot and dicot seedlings differ in the way they emerge from the soil. *(Dicot seedlings have a hook, whereas monocot seedlings grow straight out of the soil.)* Ask students to note other differences. *(The cotyledon and seed coat of a corn seed remain underground, while the cotyledons and seed coat of a bean emerge from the soil.)*

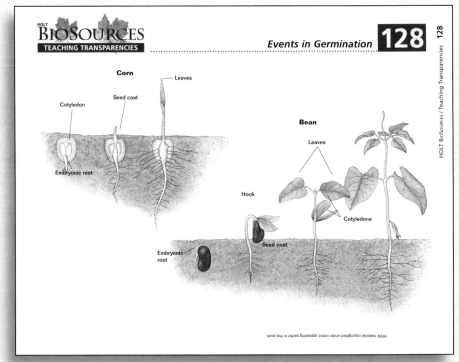

HOLT BIOSOURCES TEACHING TRANSPARENCIES

Events in Germination **128**

Corn — Leaves, Cotyledon, Seed coat, Embryonic root

Bean — Leaves, Hook, Cotyledons, Seed coat, Embryonic root

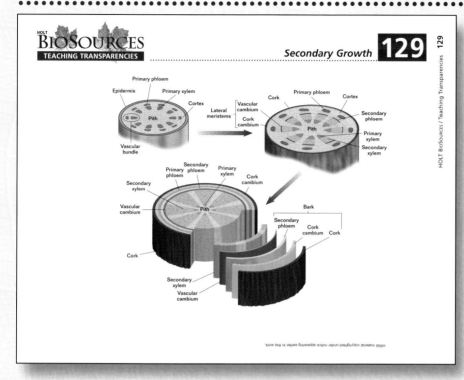

HOLT BIOSOURCES
TEACHING TRANSPARENCIES

Secondary Growth **129**

Teaching Strategies

- Use this transparency when discussing secondary growth. Review the terms *xylem, phloem, meristem,* and *vascular cambium.*
- Proceed through these diagrams showing a woody stem's structure, and help students locate each type of tissue. Point out that the vascular bundles seen in a young woody stem fuse into solid cylinders of tissue in a mature woody stem.
- Have students locate the vascular cambium and identify the type of tissue that makes up most of the inside of a mature woody stem *(secondary xylem).*

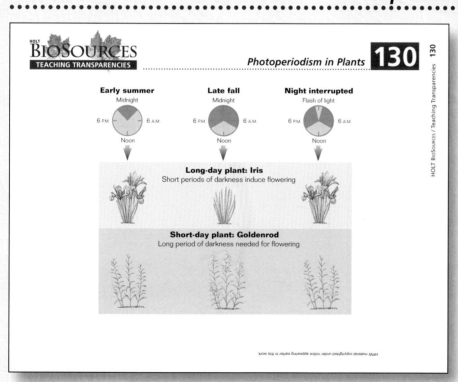

Teaching Strategies

- Use this transparency when discussing the effects of day length on flowering. Review the term *photoperiodism.*
- Proceed through the diagram. Point out that, as the third column illustrates, night length actually triggers flowering in short-day and long-day plants.
- Ask students to explain how they could cause a long-day plant to flower in the late fall. *(Interrupt each night period with a bright flash of light to simulate short nights.)*
- Ask students to describe a plant that flowers steadily through the spring, summer, and fall *(a day-neutral plant).*

Teaching Strategies

- Use this transparency when describing the basic structure of a flower. Review the terms *stamen, anther, pistil,* and *ovary.*

- Point out the parts of a typical flower that make up each floral whorl. Ask students to suggest a function of a whorl based on its appearance.

- Have students suggest ways that floral structure could be modified for particular functions, e.g., wind pollination, self-pollination, cross-pollination, and wind resistance.

- Inform students that the outermost whorl of a flower is called the calyx and that the innermost whorl is called the gynoecium. Ask them if petals are an advantage for wind pollination.

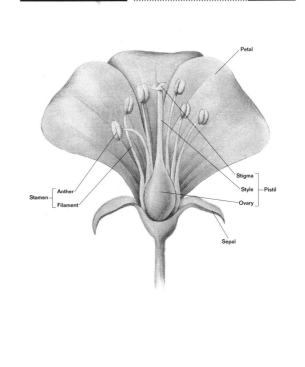

HRW material copyrighted under notice appearing earlier in this work.

Labels: Petal, Stigma, Style, Pistil, Stamen, Anther, Filament, Ovary, Sepal

132 *Cross Section of an Angiosperm Ovary*

Teaching Strategies

- Use this transparency when explaining the events of double fertilization. Review the terms *pistil, pollen tube,* and *ovule.*

- Proceed through the steps of double fertilization, beginning with a pollen grain falling on a stigma. Have students describe how the sperm travel from the pollen grain to the ovule. *(The sperm travel through a pollen tube that grows through the pistil's tissues to the ovule.)*

- Point out that the ovule contains the egg cell. One sperm fuses with this cell to form the zygote. Have students describe the fate of the second sperm.

How many sperm does a pollen grain contain?

(two)

What kind of tissue does the second sperm help form?

(a nutrient-rich tissue that will feed the embryonic plant as it develops within the seed)

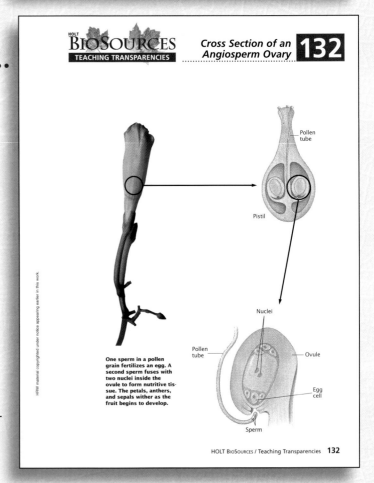

HRW material copyrighted under notice appearing earlier in this work.

One sperm in a pollen grain fertilizes an egg. A second sperm fuses with two nuclei inside the ovule to form nutritive tissue. The petals, anthers, and sepals wither as the fruit begins to develop.

Labels: Pollen tube, Pistil, Nuclei, Pollen tube, Ovule, Egg cell, Sperm

Life Cycle of a Conifer 133

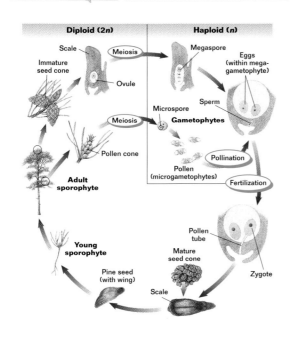

HOLT BIOSOURCES
TEACHING TRANSPARENCIES

Life Cycle of a Conifer **133**

Diploid (2n) | **Haploid (n)**

Scale
Meiosis
Immature seed cone
Ovule
Megaspore
Eggs (within mega-gametophyte)
Sperm
Microspore
Meiosis
Gametophytes
Pollen cone
Pollination
Pollen (microgametophytes)
Fertilization
Adult sporophyte
Pollen tube
Young sporophyte
Mature seed cone
Pine seed (with wing)
Zygote
Scale

HFW material copyrighted under notice appearing earlier in this work.

HOLT BIOSOURCES / Teaching Transparencies **133**

Life Cycle of a Conifer 133

Teaching Strategies

- Use this transparency when discussing the life cycle of conifers. Review the terms *spore, gametophyte, pollination,* and *sporophyte.*
- Proceed through the gymnosperm life cycle, starting with the mature sporophyte. Point out that when meiosis occurs, the chromosome number changes from diploid to haploid. Also point out that microspores and megaspores form within the male and female cones, respectively. These spores become microgametophytes (pollen grains) and megagametophytes (containing eggs), which are haploid.

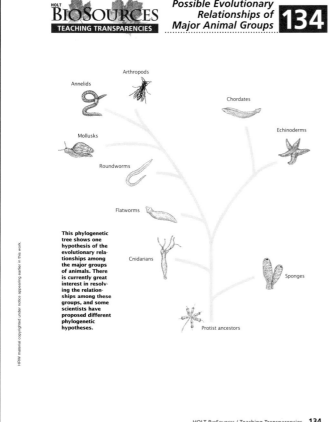

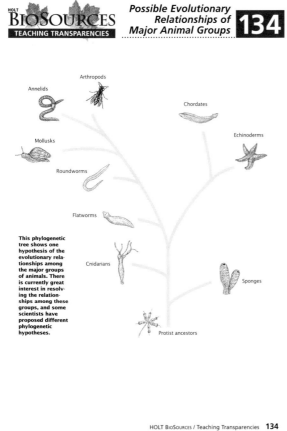

HOLT BIOSOURCES
TEACHING TRANSPARENCIES

Possible Evolutionary Relationships of Major Animal Groups **134**

Arthropods
Annelids
Chordates
Mollusks
Echinoderms
Roundworms
Flatworms
Cnidarians
Sponges
Protist ancestors

This phylogenetic tree shows one hypothesis of the evolutionary relationships among the major groups of animals. There is currently great interest in resolving the relationships among these groups, and some scientists have proposed different phylogenetic hypotheses.

HFW material copyrighted under notice appearing earlier in this work.

HOLT BIOSOURCES / Teaching Transparencies **134**

| *Animals* | Unit 7 |

Possible Evolutionary Relationships of 134 Major Animal Groups

Teaching Strategies

- Use this transparency when discussing possible evolutionary relationships among the major animal groups. Review the terms *phylogeny* and *phylogenetic tree.*
- Review the relationships depicted in the figure, starting with the ancestral protist at the bottom of the tree. Challenge students to hypothesize about the evolutionary milestone that occurs in each major animal group.
- Have students determine the oldest group of animals *(sponges)* and the groups that evolved most recently *(arthropods).*
- Have students copy this tree onto a sheet of paper, and ask them to add the following animals in their appropriate animal groups on the tree: clams *(with mollusks),* grasshoppers *(with arthropods),* mice *(with chordates),* and humans *(with chordates).*

Teaching Strategies

- Use this transparency when discussing the evolutionary relationships among arthropods.

- Remind students that mandibulates are arthropods with jaws and chelicerates are arthropods with fangs or pincers. With over 700,000 named species of insects that are mandibulates, there are many more kinds of mandibulate arthropods on the Earth than chelicerate arthropods.

- Ask students which two of the following arthropods are the most closely related: a ladybug, a pill bug, a horseshoe crab, and a millipede *(the millipede and the ladybug)*.

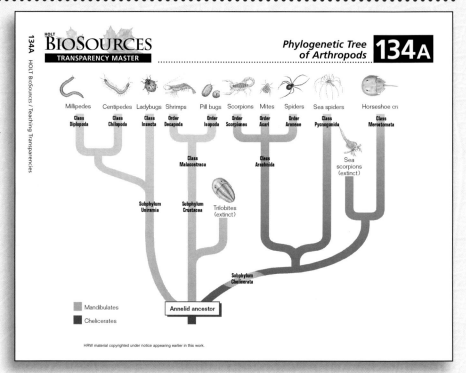

HOLT BIOSOURCES
TRANSPARENCY MASTER

Phylogenetic Tree of Arthropods 134A

134A   HOLT BioSources / Teaching Transparencies

Millipedes — Class Diplopoda
Centipedes — Class Chilopoda
Ladybugs — Class Insecta
Shrimps — Order Decapoda
Pill bugs — Order Isopoda
Scorpions — Order Scorpiones
Mites — Order Acari
Spiders — Order Araneae
Sea spiders — Class Pycnogonida
Horseshoe crab — Class Merostomata

Class Malacostraca
Class Arachnida
Sea scorpions (extinct)
Subphylum Uniramia
Subphylum Crustacea
Trilobites (extinct)
Subphylum Chelicerata

Mandibulates
Chelicerates
Annelid ancestor

HRW material copyrighted under notice appearing earlier in this work.

Teaching Strategies

- Use this transparency to summarize the evolutionary changes in the animal body. Review the terms *multicellularity, symmetry, cephalization, coelom,* and *notochord*.

- Review the characteristics of the nine stages of animal body evolution, asking students to explain why each characteristic is referred to as a milestone.

In which stage does bilateral symmetry first appear? What is a representative organism?
(in Stage 3; the liver fluke)

How does the earthworm differ from its predecessors?
(It has a segmented body.)

What does the lancelet have in common with animals such as humans and birds?
(They all have a notochord.)

HOLT BIOSOURCES
TEACHING TRANSPARENCIES

The Animal Body: An Evolutionary Journey 135

The Animal Body: An Evolutionary Journey

Phylum	Key features	Typical organism
Porifera	Multicellularity	Sponge
Cnidaria	Radial symmetry Extracellular digestion Specialized tissues	Hydra
Platyhelminthes	Internal organs Bilateral symmetry Cephalization	Liver fluke
Nematoda	Body cavity	Nematode
Mollusca	Coelom	Snail
Annelida	Segmentation	Earthworm
Arthropoda	Jointed appendages Exoskeleton	Wasp
Echinodermata	Deuterostome development Endoskeleton	Sea star
Chordata	Notochord	Lancelet

HRW material copyrighted under notice appearing earlier in this work.

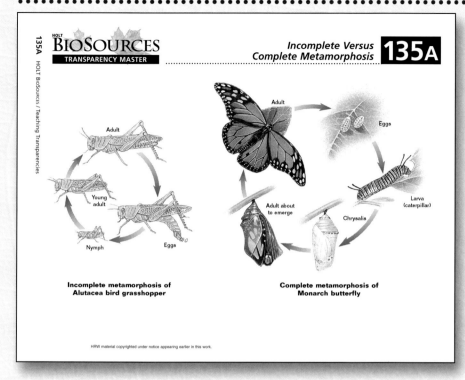

135A

HOLT BioSources / Teaching Transparencies

HOLT
BIOSOURCES
TRANSPARENCY MASTER

Incomplete Versus
Complete Metamorphosis 135A

Adult

Adult

Young
adult

Nymph   Eggs

**Incomplete metamorphosis of
Alutacea bird grasshopper**

Eggs

Larva
(caterpillar)

Adult about
to emerge

Chrysalis

**Complete metamorphosis of
Monarch butterfly**

HRW material copyrighted under notice appearing earlier in this work.

Teaching Strategies

- Use this transparency to compare incomplete metamorphosis with complete metamorphosis.
- Tell students that there are a greater number of insect species that undergo complete metamorphosis than there are insect species that undergo incomplete metamorphosis. Remind them that complete metamorphosis offers insects an evolutionary advantage because the larvae exploit different habitats and food sources than the adult.
- Have students state the common name for a butterfly or moth larva *(caterpillar)*.

Major Orders of Insects 136

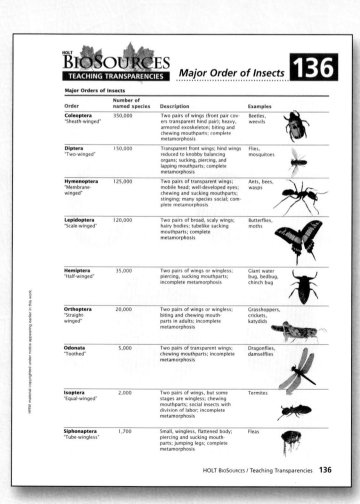

BIOSOURCES
TEACHING TRANSPARENCIES   *Major Order of Insects* 136

Major Orders of Insects

Order	Number of named species	Description	Examples	
Coleoptera "Sheath-winged"	350,000	Two pairs of wings (front pair covers transparent hind pair); heavy, armored exoskeleton; biting and chewing mouthparts; complete metamorphosis	Beetles, weevils	
Diptera "Two-winged"	150,000	Transparent front wings; hind wings reduced to knobby balancing organs; sucking, piercing, and lapping mouthparts; complete metamorphosis	Flies, mosquitoes	
Hymenoptera "Membrane-winged"	125,000	Two pairs of transparent wings; mobile head; well-developed eyes; chewing and sucking mouthparts; stinging; many species social; complete metamorphosis	Ants, bees, wasps	
Lepidoptera "Scale-winged"	120,000	Two pairs of broad, scaly wings; hairy bodies; tubelike sucking mouthparts; complete metamorphosis	Butterflies, moths	
Hemiptera "Half-winged"	35,000	Two pairs of wings or wingless; piercing, sucking mouthparts; incomplete metamorphosis	Giant water bug, bedbug, chinch bug	
Orthoptera "Straight-winged"	20,000	Two pairs of wings or wingless; biting and chewing mouthparts in adults; incomplete metamorphosis	Grasshoppers, crickets, katydids	
Odonata "Toothed"	5,000	Two pairs of transparent wings; chewing mouthparts; incomplete metamorphosis	Dragonflies, damselflies	
Isoptera "Equal-winged"	2,000	Two pairs of wings, but some stages are wingless; chewing mouthparts; social insects with division of labor; incomplete metamorphosis	Termites	
Siphonaptera "Tube-wingless"	1,700	Small, wingless, flattened body; piercing and sucking mouthparts; jumping legs; complete metamorphosis	Fleas	

HOLT BioSources / Teaching Transparencies **136**

Teaching Strategies

- Use this transparency to summarize the main characteristics of the major insect orders. Review the terms *exoskeleton, endoskeleton, complete metamorphosis,* and *incomplete metamorphosis.*
- Have students identify the order that contains the largest number of species *(order Coleoptera)*. Inform students that there are more kinds of beetles than all the kinds of non-insect animals put together. Also remind students that this table depicts only some of the larger insect orders. There are about 30 orders of insects currently recognized.
- Ask students to describe how insects grow in size. *(They shed and discard their exoskeletons periodically through the process of molting.)*

Teaching Strategies

- Use this transparency to illustrate the structure of the marine worm Nereis. Review the terms *annelid* and *polychaete*.

- Remind students that an annelid is classified according to the kind of external appendages it has. The parapodia that *Nereis* has indicate that this annelid is a marine polychaete.

- Have students predict how *Nereis* feeds, based on its appearance. *(The presence of jaws indicates that Nereis is a predator.)*

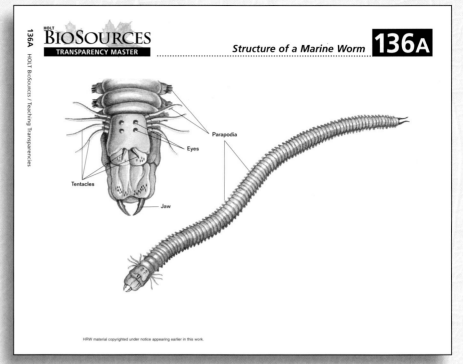

136A HOLT BioSources / Teaching Transparencies

HOLT
BIOSOURCES
TRANSPARENCY MASTER

Structure of a Marine Worm 136A

Parapodia

Eyes

Tentacles

Jaw

Teaching Strategies

- Use this transparency when describing the life cycle of the monarch butterfly. Review the terms *complete metamorphosis, larva*, and *chrysalis*.

- Point out that the organism commonly known as a caterpillar is actually the larval stage of the butterfly. Have students explain why complete metamorphosis offers insects an evolutionary advantage *(because the larvae exploit different habitats and food sources than the adult).*

- Remind students that 90 percent of all insect species have a life cycle that involves complete metamorphosis.

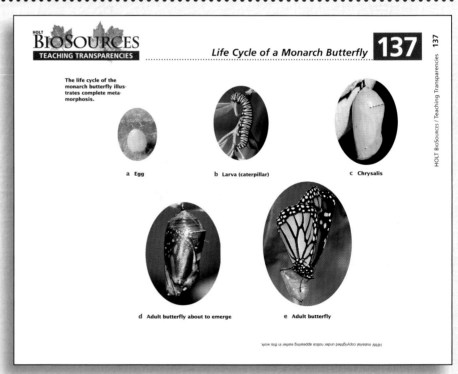

HOLT
BIOSOURCES
TEACHING TRANSPARENCIES

Life Cycle of a Monarch Butterfly 137

137 HOLT BioSources / Teaching Transparencies

The life cycle of the monarch butterfly illustrates complete metamorphosis.

a Egg

b Larva (caterpillar)

c Chrysalis

d Adult butterfly about to emerge

e Adult butterfly

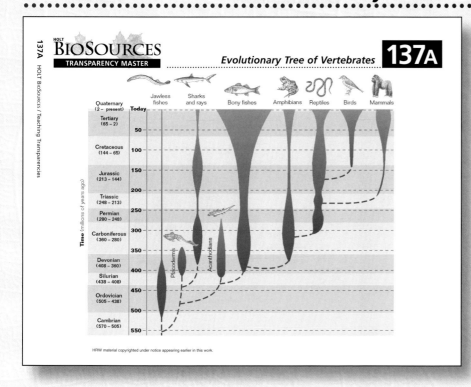

Teaching Strategies

- Use this transparency when discussing the evolutionary relationships among vertebrates.

- Point out that this diagram shows the waxing and waning of diversity in each of the major classes of vertebrates. Students should be able to correlate the major mass extinction with declines in diversity in major groups. Thus, the mass extinction at the end of the Permian period is reflected in reductions in diversity in sharks and rays, bony fishes, amphibians, and reptiles.

- Ask students which group of vertebrates evolved most recently *(birds)*.

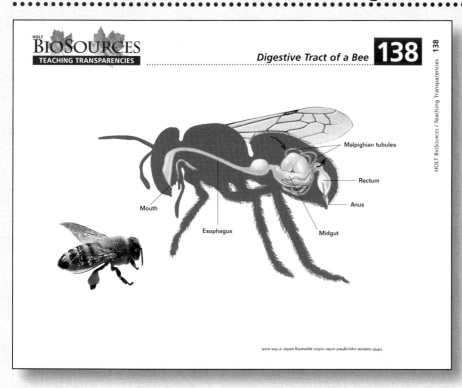

Teaching Strategies

- Use this transparency when describing the anatomy of a bee. Review the terms *esophagus* and *Malpighian tubules*.

- Have students locate the Malpighian tubules that extend from the bee's gut. Explain that a mixture of wastes and useful molecules passes into the tubules from the surrounding blood that bathes them. As this mixture travels down the gut, the useful molecules are reabsorbed into the blood. The wastes remain in the gut and are excreted.

- Ask students to name the kinds of useful molecules that a bee would reabsorb into its body from its gut.

Swim Bladder in Bony Fish

Teaching Strategies

- Use this transparency to illustrate the evolution of the swim bladder in fish.
- Point out to students that the swim bladder formed as an outpocketing of the pharynx, the portion of the throat that leads to the digestive tract. Show them that in modern bony fishes, the swim bladder has become an independent organ, completely separate from the pharynx.
- Have students describe the function of the swim bladder. *(It enables a fish to rise or sink in the water without having to swim.)*

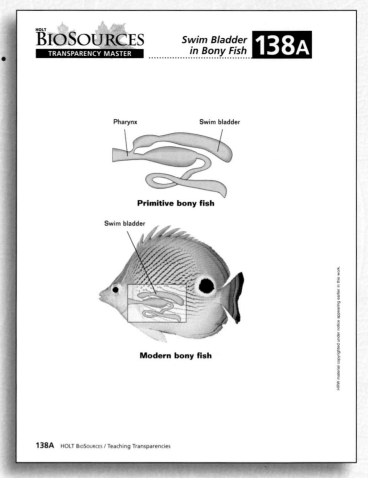

HOLT
BIOSOURCES
TRANSPARENCY MASTER

Swim Bladder
in Bony Fish **138A**

Pharynx   Swim bladder

Primitive bony fish

Swim bladder

Modern bony fish

138A   HOLT BIOSOURCES / Teaching Transparencies

Tracheal System of a Beetle

Teaching Strategies

- Use this transparency when describing the anatomy of a beetle. Review the terms *trachea* and *spiracles*.
- Compare and contrast the tracheal system of a beetle with the trachea of a human. *(The structures are similar in that they function to bring outside air into the organism. They are different in that the beetle's tracheal system has many openings to the outside, while the human trachea has just one.)*
- Ask students to discuss possible disadvantages of having many openings to the respiratory system. *(Students may state that there is an increased chance of infection and water loss.)*

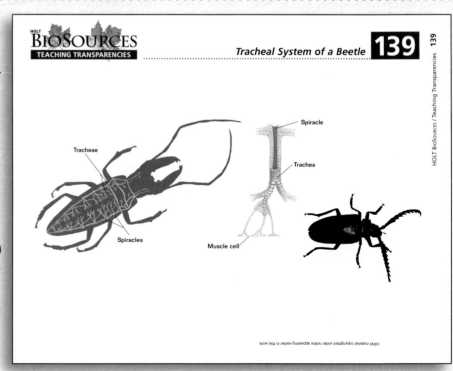

HOLT
BIOSOURCES
TEACHING TRANSPARENCIES

Tracheal System of a Beetle **139**

Tracheae

Spiracle

Trachea

Spiracles

Muscle cell

HOLT BIOSOURCES / Teaching Transparencies   139

Lateral Line in Bony Fish 139A

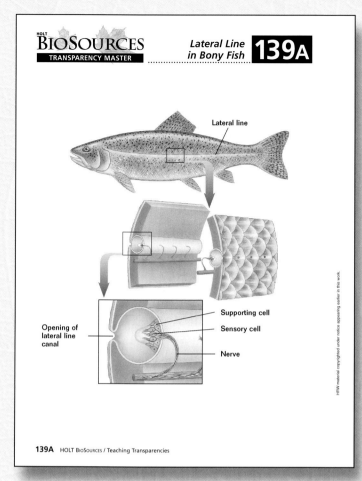

BioSources
TRANSPARENCY MASTER

Lateral Line in Bony Fish **139A**

Lateral line

Supporting cell
Sensory cell
Opening of lateral line canal
Nerve

139A HOLT BioSources / Teaching Transparencies

Lateral Line in Bony Fish 139A

Teaching Strategies

- Use this transparency when discussing the lateral line in a bony fish.
- Tell students that the lateral line is an important sense organ in a bony fish. The sensory cells located inside a lateral line enable a fish to determine the speed and direction of the water current and the position of objects in the water.
- Have students name a structure in the human body that functions in a way similar to that of a fish's lateral line *(the inner ear).*

BioSources
TEACHING TRANSPARENCIES

Insect Diversity **140**

Fungi
Flowering plants
Beetles
Protists
Other animals
Mollusks
Bees, wasps
Vertebrates
Butterflies, moths
Spiders
Other insects
Flies, mosquitoes
Millipedes, centipedes

The predominance of insects in the living world is demonstrated by the blue sections of this pie chart. Notice that there are more beetles than any other kind of insect.

HOLT BioSources / Teaching Transparencies **140**

Insect Diversity 140

Teaching Strategies

- Use this transparency to summarize the range of insect diversity.
- Tell students that although the current geological time period is sometimes called the Age of Mammals, this pie chart clearly shows that insects, not mammals, are the most abundant group of animals on Earth today.
- Have students name the arthropods, other than insects, that are represented in the pie chart *(millipedes, centipedes, and spiders).*

Teaching Strategies

- Use this transparency when discussing the structure of the opercula in fish.
- Remind students that the movements of a bony fish's opercula pump water over the gills, allowing the fish to breathe even when it is stationary. Less-advanced fishes, like sharks, have to swim forward with their mouths open to move water over their gills.
- Ask students to describe what happens when water moves over gills. *(Oxygen from the water moves into the gills, and carbon dioxide from the gills moves into the water.)*

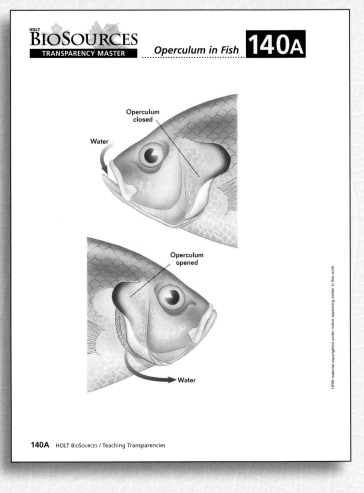

Operculum in Fish **140A**

Operculum closed

Water

Operculum opened

Water

140A HOLT BioSources / Teaching Transparencies

Teaching Strategies

- Use this transparency to stimulate a discussion on the anatomy of the clam. Review the terms *gill, foot,* and *siphon*.
- Review the internal structures of the clam, reminding students that the clam represents the basic mollusk body plan. Inform students that the mollusk body consists of four principal parts: a head region, a muscular foot, a visceral mass, and a mantle. Have students identify these regions in the illustration.
- Have students describe how a mollusk moves across surfaces in its environment. *(Most mollusks creep along slowly on a flattened muscular foot.)*

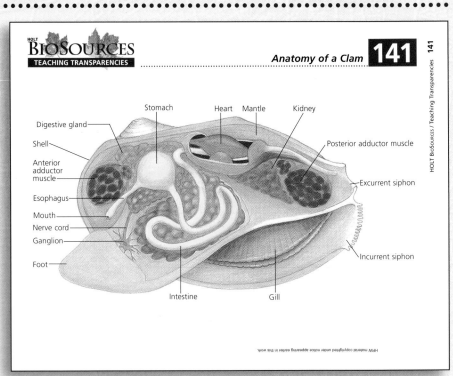

Anatomy of a Clam **141**

Stomach   Heart   Mantle   Kidney

Digestive gland

Shell

Anterior adductor muscle

Esophagus

Mouth

Nerve cord

Ganglion

Foot

Posterior adductor muscle

Excurrent siphon

Incurrent siphon

Intestine   Gill

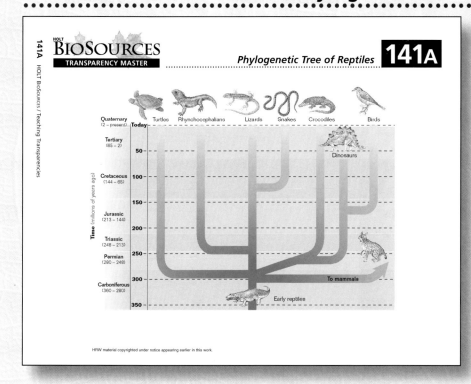

Phylogenetic Tree of Reptiles **141A**

Teaching Strategies

- Use this transparency to illustrate the evolutionary relationships among the groups of living reptiles.
- Point out that, out of all the modern reptiles, turtles are thought to be the most closely related to the early reptiles. Also point out the close relationship between lizards and snakes (snakes evolved from lizards) and between crocodiles and birds (both groups share a common ancestor).
- Ask students which group of reptiles contains the immediate ancestors of birds *(dinosaurs)*.

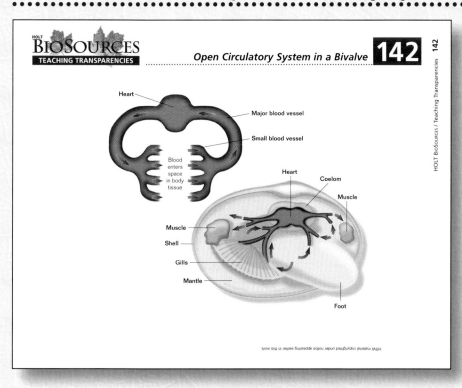

Open Circulatory System in a Bivalve **142**

Teaching Strategies

- Use this transparency when discussing the open circulatory system of mollusks. Review the terms *mantle cavity* and *gills*.
- Explain to students that after blood leaks out of the open circulatory system of a mollusk, the blood bathes the mollusk's tissues. Nutrients, oxygen, and carbon dioxide are exchanged between the blood and the tissues.
- Ask students to explain why a bivalve must take water into its mantle cavity. *(The gills, the organs that extract oxygen from the water, are located in the mantle cavity.)*

142A Most Likely Cause of Dinosaur Extinction

Teaching Strategies

- Use this transparency to illustrate the most widely accepted explanation of the extinction of the dinosaurs.
- Inform students that a large meteorite or comet struck the Earth on what is now the Yucatan peninsula, leaving a crater 300 km (185 mi.) in diameter. Such an impact would have produced a huge cloud of debris that could have blocked out sunlight and caused a worldwide period of cool temperatures.
- Have students explain why mammals could have survived a period of cool temperatures. *(They are endothermic and insulated by fur.)*

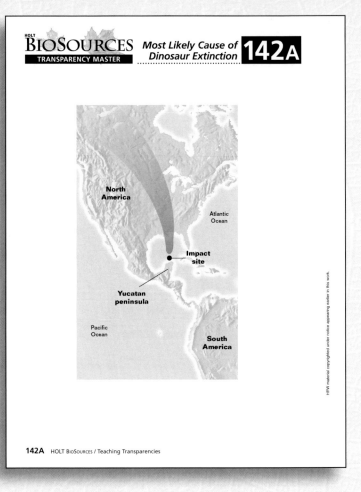

BioSources TRANSPARENCY MASTER — *Most Likely Cause of Dinosaur Extinction* **142A**

North America

Atlantic Ocean

Impact site

Yucatan peninsula

Pacific Ocean

South America

142A HOLT BioSources / Teaching Transparencies

---

143 Three Types of Body Construction

Teaching Strategies

- Use this transparency to discuss evolution of a body cavity. Review the terms *acoelomate, pseudocoelomate,* and *coelomate.*
- Review each body plan, beginning with the acoelomates. Explain that without the development of a body cavity there is no organ system development. Have students explain how an acoelomate, such as a flatworm, gets the oxygen it needs to live. *(Oxygen diffuses through the flatworm's skin.)*
- Inform students that roundworms were the first animals to develop a body cavity (pseudocoelom).
- Explain that most animals, including earthworms, develop a coelom. Describe how the coelom's structure matches its function.

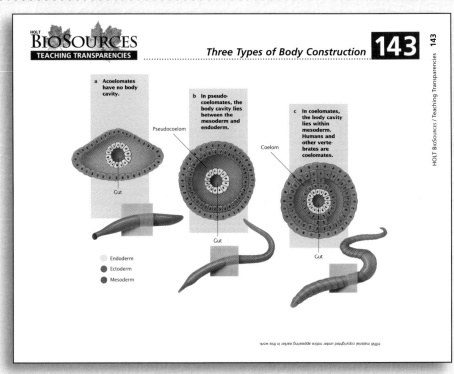

BioSources TEACHING TRANSPARENCIES — *Three Types of Body Construction* **143**

a Acoelomates have no body cavity.

b In pseudocoelomates, the body cavity lies between the mesoderm and endoderm.

Pseudocoelom

c In coelomates, the body cavity lies within mesoderm. Humans and other vertebrates are coelomates.

Coelom

Gut

Gut

Gut

Endoderm
Ectoderm
Mesoderm

Structure of a Feather 143A

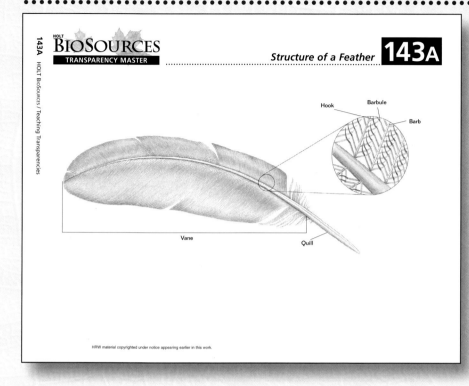

BioSources
TRANSPARENCY MASTER
Structure of a Feather 143A

Hook Barbule
Barb
Vane
Quill

HRW material copyrighted under notice appearing earlier in this work.

Teaching Strategies

- Use this transparency to illustrate how a feather keeps its shape.
- Point out the feather's hooks, which fasten neighboring barbules to one another, creating an interlocking structure. Point out that this interlocking structure helps produce an aerodynamic surface that reduces wind resistance.
- Have students explain why a bird wing can be composed of feathers, but not of hair. (*Hair lacks interlocking hooks and barbules and therefore cannot form a continuous surface.*)

Structure of a Sponge 144

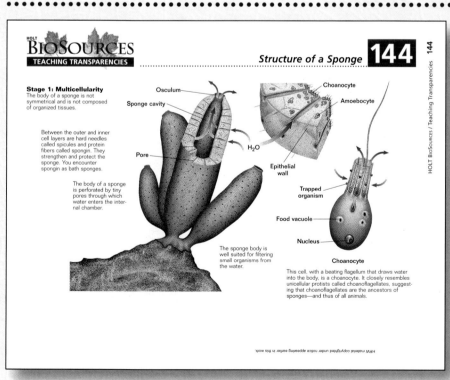

BioSources
TEACHING TRANSPARENCIES
Structure of a Sponge 144

Stage 1: Multicellularity
The body of a sponge is not symmetrical and is not composed of organized tissues.

Between the outer and inner cell layers are hard needles called spicules and protein fibers called spongin. They strengthen and protect the sponge. You encounter spongin as bath sponges.

The body of a sponge is perforated by tiny pores through which water enters the internal chamber.

The sponge body is well suited for filtering small organisms from the water.

Osculum
Sponge cavity
Pore
H_2O
Choanocyte
Amoebocyte
Epithelial wall
Trapped organism
Food vacuole
Nucleus
Choanocyte

This cell, with a beating flagellum that draws water into the body, is a choanocyte. It closely resembles unicellular protists called choanoflagellates, suggesting that choanoflagellates are the ancestors of sponges—and thus of all animals.

HRW material copyrighted under notice appearing earlier in this work.

Teaching Strategies

- Use this transparency when describing sponge morphology. Review the terms *osculum, choanocyte,* and *amoebocyte.* Remind students that sponges are sessile; they live attached to the sea bottom or some other submerged object.
- Have students explain the function of collar cells. (*Collar cells line the inside of a sponge and trap food particles.*)

How does the choanocyte take food particles into its cytoplasm?
(by endocytosis)

- Point out the proteinaceous spicules that form the skeleton of the sponge. The skeletons of bath sponges (*Spongia*) are cleaned and sold for commercial purposes.

Comparison of Coyote and Beaver Skulls

Teaching Strategies

- Use this transparency when comparing coyote skull structures with beaver skull structures.

- Point out that the coyote's teeth function like a pair of scissors; two sharp surfaces slide across one another as the coyote closes its mouth. Students should see that the coyote's teeth are unsuited for grinding plant material. The beaver's molars are suited to this task because they function like a grinding mill; two rough surfaces come together and slide across one another when the beaver closes its mouth.

- Ask students which animal's molars resemble human molars the most *(beaver)*.

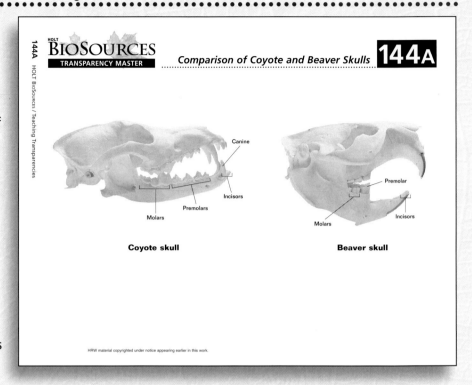

144A BIOSOURCES TRANSPARENCY MASTER *Comparison of Coyote and Beaver Skulls* **144A**

HOLT BioSources / Teaching Transparencies

Canine

Incisors

Premolars

Molars

Premolar

Molars

Incisors

Coyote skull

Beaver skull

HRW material copyrighted under notice appearing earlier in this work.

Sexual Reproduction of Sponges

Teaching Strategies

- Use this transparency when discussing sexual reproduction in sponges. Review the terms *egg, sperm,* and *larva*.

- Tell students that the presence of gametes in this figure indicates that the two sponges on the left are sexually reproducing. Ask students to state the advantage of sexual reproduction. *(It introduces genetic variation into offspring, thus increasing their chance for survival.)*

- Point out the motile larvae in the illustration. Ask students to explain the evolutionary advantage of producing larvae that can swim. *(The larvae can swim to different areas and take advantage of new resources, like food and space.)*

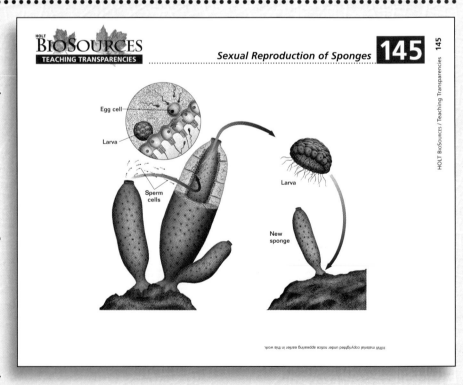

BIOSOURCES TEACHING TRANSPARENCIES *Sexual Reproduction of Sponges* **145** 145

HOLT BioSources / Teaching Transparencies

Egg cell

Larva

Sperm cells

Larva

New sponge

HRW material copyrighted under notice appearing earlier in this work.

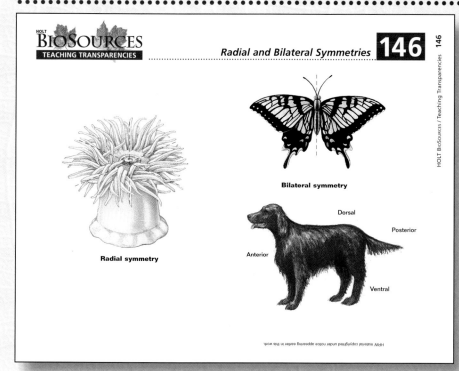

Teaching Strategies

- Use this transparency when comparing radial and bilateral symmetries. Review the terms *anterior, posterior, dorsal,* and *ventral.*

- Explain to students that almost all animals have bilateral symmetry, like the butterfly and the dog shown in the figure. Those animals that are radially symmetric are all aquatic, such as the sea anemone shown in the figure.

- Have students name an animal, other than a cnidarian or ctenophore, that has radial symmetry *(a sea star or some other echinoderm).*

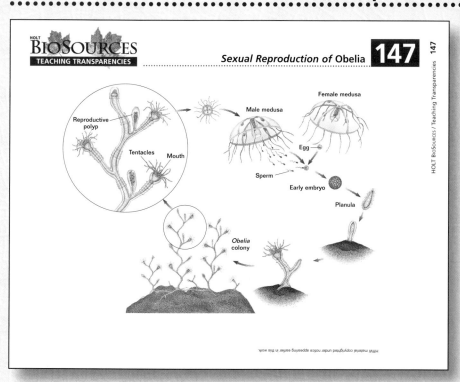

Teaching Strategies

- Use this transparency when discussing sexual reproduction in Obelia. Review the terms *polyp, medusa,* and *planula.*

- Point out that Obelia forms complex colonies that contain both feeding polyps and reproductive polyps. Male and female medusae released by the reproductive polyps sexually reproduce, resulting in free-swimming larvae called planulae. Eventually planulae settle to the bottom and grow into a new colony.

- Have students explain how the tentacles of Obelia function to capture food. *(Stinging cells located in the tentacles shoot out harpoon-like nematocysts into prey.)*

Teaching Strategies

• Use this transparency when discussing sexual reproduction in Aurelia. Review the terms *planula, polyp,* and *medusa.*

• Tell students that Aurelia, like all jellyfishes, spends most of its life as a free-floating medusa. Most jellyfish species also go through a small, inconspicuous polyp stage at some point in their life cycle, as shown in this figure.

• Have students compare the symmetry of Aurelia medusa with the symmetry of an Aurelia polyp. *(They both have radial symmetry.)*

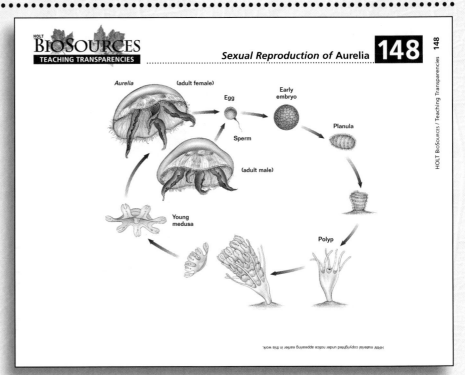

148 *Sexual Reproduction of Aurelia*

Teaching Strategies

• Use this transparency when discussing the life cycle of the blood fluke, *Schistosoma.*

• Begin with the mature fluke in the blood vessel of an intestine. Point out how the appearance of a blood fluke changes as it develops from larva to adult. Adult blood flukes grow up to an inch long and may live for years in the blood vessels of their human host.

• Ask students if they could expect to find a body cavity inside a blood fluke. *(No, because a blood fluke is a flatworm, which is an acoelomate.)*

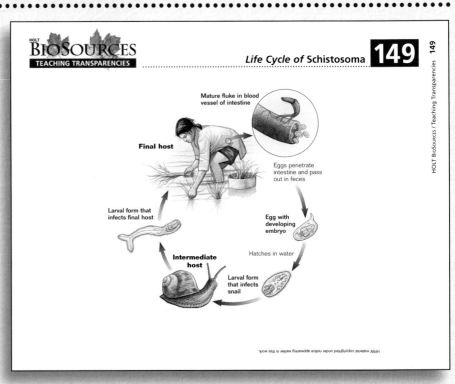

149 *Life Cycle of Schistosoma*

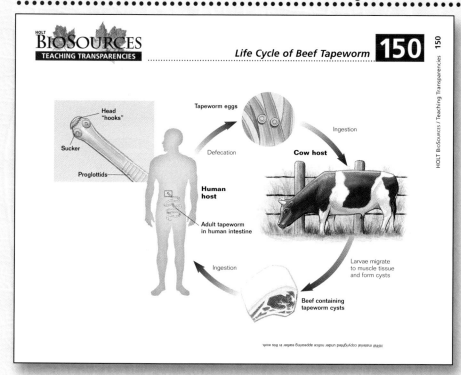

BioSources
TEACHING TRANSPARENCIES

Life Cycle of Beef Tapeworm **150**

Head "hooks"

Sucker

Proglottids

Tapeworm eggs

Ingestion

Defecation

Cow host

Human host

Adult tapeworm in human intestine

Ingestion

Larvae migrate to muscle tissue and form cysts

Beef containing tapeworm cysts

Teaching Strategies

- Use this transparency when discussing the life cycle of the beef tapeworm, *Taenia saginata*. Review the terms *cyst* and *proglottid*.

- Begin with the tapeworm eggs lying in the grass. Point out that cows become infected with tapeworms by eating grass contaminated with tapeworm eggs. The eggs make their way to the grass when human feces that contain the eggs are improperly disposed of.

- Have students explain how they can avoid getting beef tapeworm. *(Eat only beef that has been cooked to a temperature high enough to kill the larval tapeworms that are encysted in the meat.)*

Nephridium Function in Chiton 151

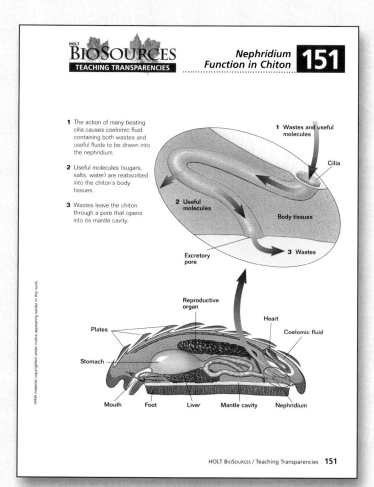

BioSources
TEACHING TRANSPARENCIES

Nephridium Function in Chiton **151**

1 The action of many beating cilia causes coelomic fluid containing both wastes and useful fluids to be drawn into the nephridium.

2 Useful molecules (sugars, salts, water) are reabsorbed into the chiton's body tissues.

3 Wastes leave the chiton through a pore that opens into its mantle cavity.

1 Wastes and useful molecules

Cilia

2 Useful molecules

Body tissues

3 Wastes

Excretory pore

Reproductive organ

Heart

Coelomic fluid

Plates

Stomach

Mouth   Foot   Liver   Mantle cavity   Nephridium

HOLT BioSources / Teaching Transparencies **151**

Teaching Strategies

- Use this transparency when discussing the excretory system in mollusks. Review the terms *coelomic fluid, nephridium,* and *mantle cavity*.

- Compare the function of a mollusk's nephridium to the function of a human kidney. Tell students that both structures have the same basic function. They conserve the useful molecules from a fluid and eliminate the wastes.

- Have students identify where nitrogen-rich wastes accumulate in the mollusk before the wastes enter the nephridia. *(The wastes accumulate in the fluid that fills the mollusk's coelom.)*

Teaching Strategies

- Use this transparency when discussing the circulatory system in annelids. Review the terms *gut, dorsal vessel*, and *ventral vessel*.

- Ask students if blood can leak out of a closed circulatory system of an earthworm *(no)*. Explain that pressure builds in the worm's circulatory system each time the heart pumps send blood to its tissues at a fast rate. The human circulatory system, also a closed system, functions in much the same way.

- Ask students to explain what has happened to an earthworm if it is found bleeding. *(There must be a hole in its circulatory system.)*

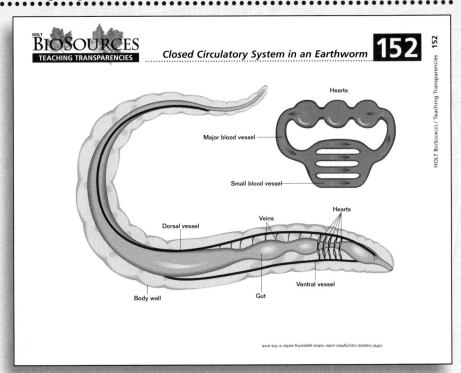

Teaching Strategies

- Use this transparency when discussing the morphology of a cnidarian. Review the term *radial symmetry*.

- Analyze the structure of Hydra. Emphasize that cnidarians are aquatic organisms that have a hollow gut with a single opening and flexible fingerlike tentacles. Have students explain how the cnidarian body is more complex than that of a sponge.

- Ask students to explain how a cnidarian captures prey. *(They capture prey with their fingerlike tentacles and paralyze the organisms with their stinging tentacles.)*

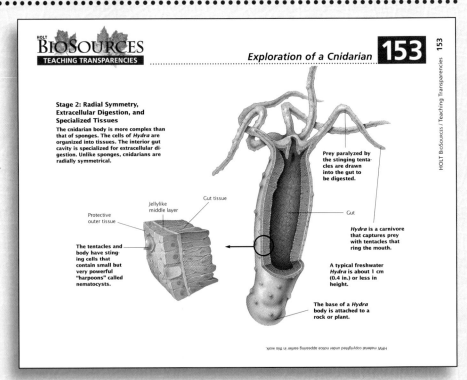

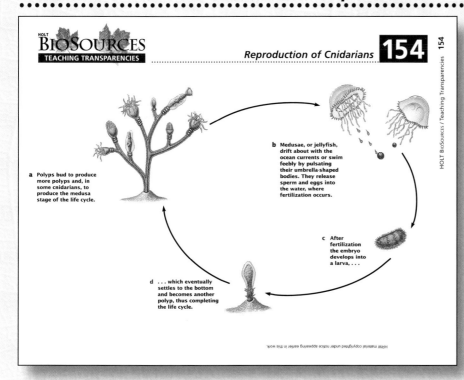

BIOSOURCES
TEACHING TRANSPARENCIES ········ *Reproduction of Cnidarians* **154**

a Polyps bud to produce more polyps and, in some cnidarians, to produce the medusa stage of the life cycle.

b Medusae, or jellyfish, drift about with the ocean currents or swim feebly by pulsating their umbrella-shaped bodies. They release sperm and eggs into the water, where fertilization occurs.

c After fertilization the embryo develops into a larva, . . .

d . . . which eventually settles to the bottom and becomes another polyp, thus completing the life cycle.

Teaching Strategies

- Use this transparency when discussing reproduction in cnidarians. Review the terms *polyp* and *medusa*.
- Lead students through the cnidarian life cycle. Emphasize that many cnidarians have two different body forms during their life cycle, polyps and medusae. Have students explain how these two forms differ. *(The polyp body form is sedentary and reproduces asexually; the medusa body form is free-swimming and reproduces sexually.)*
- Ask students to name the kinds of gametes jellyfish produce during sexual reproduction *(eggs and sperm)*.

Development of a Cnidarian Embryo 155

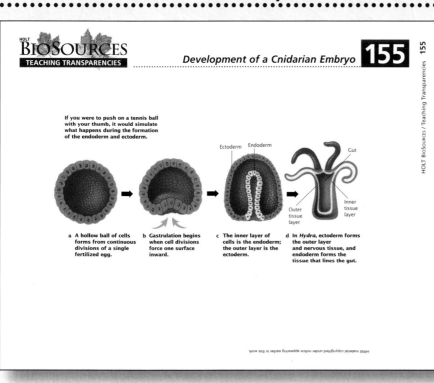

BIOSOURCES
TEACHING TRANSPARENCIES ········ *Development of a Cnidarian Embryo* **155**

If you were to push on a tennis ball with your thumb, it would simulate what happens during the formation of the endoderm and ectoderm.

Ectoderm Endoderm

Gut

Outer tissue layer

Inner tissue layer

a A hollow ball of cells forms from continuous divisions of a single fertilized egg.

b Gastrulation begins when cell divisions force one surface inward.

c The inner layer of cells is the endoderm; the outer layer is the ectoderm.

d In *Hydra*, ectoderm forms the outer layer and nervous tissue, and endoderm forms the tissue that lines the gut.

Teaching Strategies

- Use this transparency when discussing the process of gastrulation in a cnidarian embryo. Review the terms *gastrulation, endoderm*, and *ectoderm*.
- Proceed through the development of the cnidarian embryo. Be sure that students can distinguish the layer of endoderm tissue from that of ectoderm tissue. Have students design a graphic organizer that illustrates the fates of these two types of tissue layers.
- Emphasize that the empty space in the hollow ball of cells becomes the cavity of the gut. Have students write a paragraph that describes the process of gastrulation in their own words.

156 Development of a Flatworm Embryo

Teaching Strategies

- Use this transparency when discussing the process of gastrulation in a flatworm embryo. Review the terms *endoderm, ectoderm,* and *mesoderm.*

- Proceed through the development of the flatworm embryo. Be sure that students can identify the endoderm, mesoderm, and ectoderm tissue layers. Have students design a graphic organizer that depicts the fates of these three types of tissue.

- Have students write a paragraph that compares the process of gastrulation in a flatworm embryo with that in a cnidarian embryo.

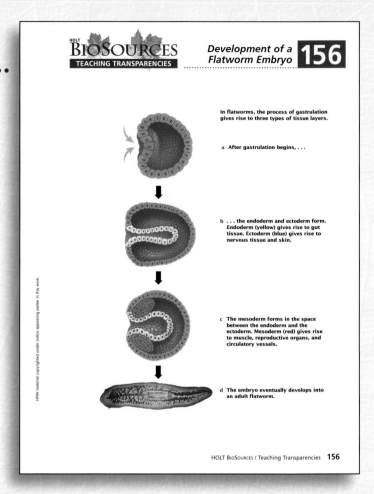

In flatworms, the process of gastrulation gives rise to three types of tissue layers.

a After gastrulation begins, . . .

b . . . the endoderm and ectoderm form. Endoderm (yellow) gives rise to gut tissue. Ectoderm (blue) gives rise to nervous tissue and skin.

c The mesoderm forms in the space between the endoderm and the ectoderm. Mesoderm (red) gives rise to muscle, reproductive organs, and circulatory vessels.

d The embryo eventually develops into an adult flatworm.

HRW material copyrighted under notice appearing earlier in this work.

157 Exploration of a Flatworm

Teaching Strategies

- Use this transparency when describing the structure and main characteristics of a flatworm. Review the terms *bilateral symmetry* and *cephalization.*

- Point out that flatworms are acoelomates, which means that they lack a body cavity. Also, point out the distinct head region. Have students describe how the liver fluke's head is modified for burrowing.

- Have students explain digestion and elimination in the flatworm. *(A flatworm has a two-way gut, which means it consumes food and eliminates waste through the same opening.)*

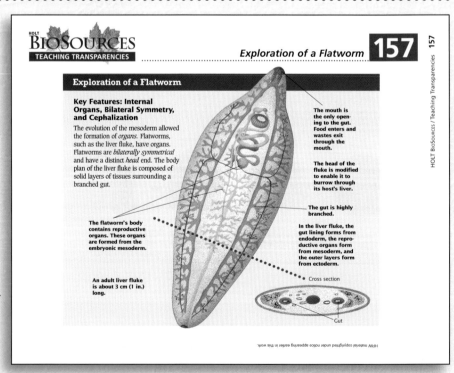

Exploration of a Flatworm

Key Features: Internal Organs, Bilateral Symmetry, and Cephalization

The evolution of the mesoderm allowed the formation of *organs.* Flatworms, such as the liver fluke, have organs. Flatworms are *bilaterally symmetrical* and have a distinct *head* end. The body plan of the liver fluke is composed of solid layers of tissues surrounding a branched gut.

The mouth is the only opening to the gut. Food enters and wastes exit through the mouth.

The head of the fluke is modified to enable it to burrow through its host's liver.

The gut is highly branched.

In the liver fluke, the gut lining forms from endoderm, the reproductive organs form from mesoderm, and the outer layers form from ectoderm.

The flatworm's body contains reproductive organs. These organs are formed from the embryonic mesoderm.

An adult liver fluke is about 3 cm (1 in.) long.

Cross section

Gut

HRW material copyrighted under notice appearing earlier in this work.

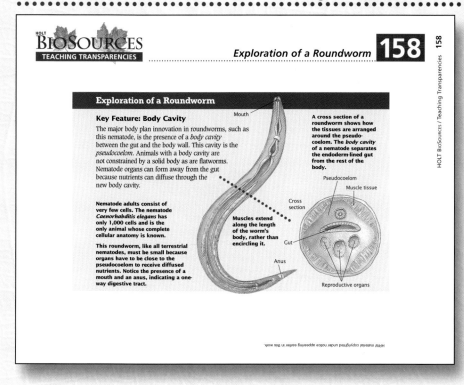

Teaching Strategies

- Use this transparency when describing the characteristics of a roundworm. Review the term *pseudocoelom*.

- Point out that roundworms are pseudocoelomates, which means they have a body cavity between the gut and the body wall. Have students explain how the development of a pseudocoelom affected organ placement in the nematode body. *(Organs could form in the body cavity, away from the gut.)*

- Have students compare the process of digestion in a roundworm with that in a flatworm. *(A roundworm has a one-way digestive tract, which is more efficient than the flatworm's two-way digestive process.)*

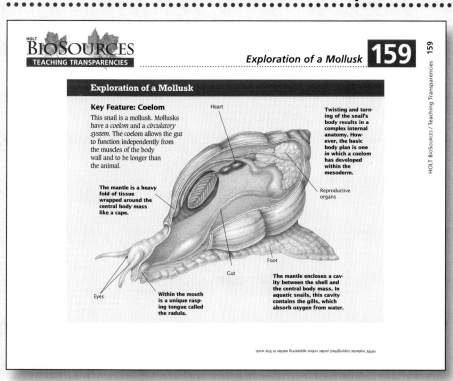

Teaching Strategies

- Use this transparency when describing the characteristics of a mollusk, such as a snail. Review the terms *coelom, foot,* and *mantle.*

- Point out that mollusks are coelomates, which means they have a fluid-filled mesoderm. Point out that the coelom separates the muscles of the body wall from the muscles that surround the gut.

- Emphasize that a coelom and a circulatory system first evolved in the phylum Mollusca, which contains animals such as snails, clams, squids, and mussels.

Teaching Strategies

- Use this transparency when reviewing the characteristics of an annelid, such as an earthworm. Review the terms *segmentation* and *setae*.

- Point out the repeating body segments. Have students explain how the earthworm's organ systems are distributed throughout its segments. *(The parts of the excretory, circulatory, and nervous systems are repeated in each segment.)*

- Have students write a paragraph that describes the advantages of having a body composed of segments. *(Segmentation offers evolutionary flexibility.)*

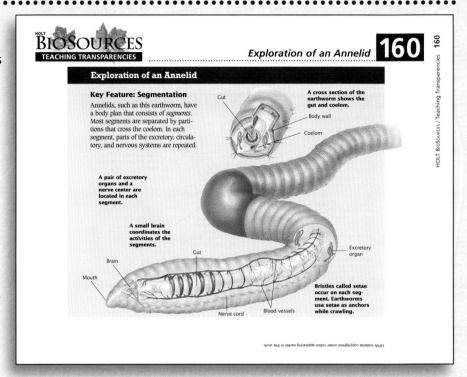

BIOSOURCES
TEACHING TRANSPARENCIES

Exploration of an Annelid **160**

Exploration of an Annelid

Key Feature: Segmentation

Annelids, such as this earthworm, have a body plan that consists of *segments*. Most segments are separated by partitions that cross the coelom. In each segment, parts of the excretory, circulatory, and nervous systems are repeated.

A cross section of the earthworm shows the gut and coelom.

Gut
Body wall
Coelom

A pair of excretory organs and a nerve center are located in each segment.

A small brain coordinates the activities of the segments.

Brain
Mouth
Gut
Excretory organ

Nerve cord
Blood vessels

Bristles called setae occur on each segment. Earthworms use setae as anchors while crawling.

Teaching Strategies

- Use this transparency when reviewing the characteristics of arthropods. Review the terms *exoskeleton* and *thorax*.

- Point out the jointed appendages in the wasp in the illustration. Have students explain the evolutionary advantage of having a body composed of jointed appendages.

- Ask students to describe the function of an arthropod's exoskeleton.

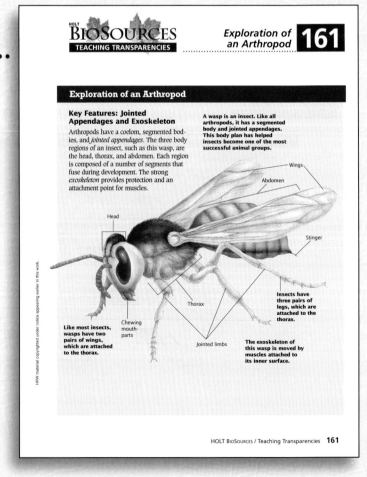

BIOSOURCES
TEACHING TRANSPARENCIES

Exploration of an Arthropod **161**

Exploration of an Arthropod

Key Features: Jointed Appendages and Exoskeleton

Arthropods have a coelom, segmented bodies, and *jointed appendages*. The three body regions of an insect, such as this wasp, are the head, thorax, and abdomen. Each region is composed of a number of segments that fuse during development. The strong *exoskeleton* provides protection and an attachment point for muscles.

A wasp is an insect. Like all arthropods, it has a segmented body and jointed appendages. This body plan has helped insects become one of the most successful animal groups.

Wings
Abdomen

Head

Stinger

Insects have three pairs of legs, which are attached to the thorax.

Thorax

Like most insects, wasps have two pairs of wings, which are attached to the thorax.

Chewing mouthparts

Jointed limbs

The exoskeleton of this wasp is moved by muscles attached to its inner surface.

HOLT BIOSOURCES / Teaching Transparencies **161**

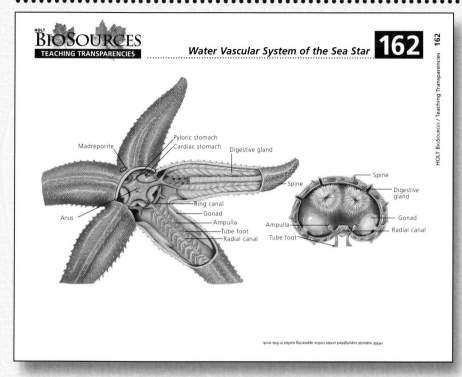

Teaching Strategies

- Use this transparency when describing the unique water vascular system in echinoderms. Review the terms *radial canal, ring canal,* and *tube foot.*

- Explain that the water vascular system is a system of fluid-filled tubes, or canals. Water enters the system through a sievelike plate on the surface of the animal. A tube from this plate leads to a series of canals that run throughout the body.

- Point out the rows of tube feet that project out from the sea star's arms. Ask students to explain how a sea star moves. *(It moves about slowly by extending its tube feet and attaching the suction cups to a solid surface.)*

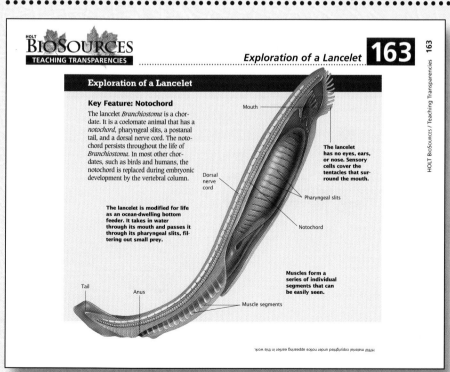

Teaching Strategies

- Use this transparency when reviewing the characteristics of a lancelet. Review the term *notochord.*

- Inform students that lancelets spend most of their time in shallow water buried tail down in the sand, with only their anterior end exposed.

- Have students explain the evolutionary advantage of having a body that has muscles and a notochord. *(Such features permit more efficient locomotion.)*

Teaching Strategies

- Use this transparency when discussing the evolution of jaws in fishes. Review the terms *gill arch* and *gill slit*.
- Tell students that the evolution of jaws in fishes played a major role in their ability to become successful predators. Jaws first developed about 440 million years ago, in a group of fishes called acanthodians.
- Have students name the structures from which jaws developed *(gill arches)*.

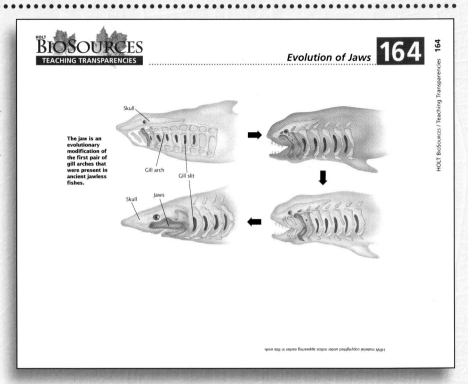

Teaching Strategies

- Use this transparency when discussing the anatomy of a bony fish. Review the term *gas bladder*. Inform students that this structure is also called a swim bladder.
- Review the internal structures of a bony fish.
- Explain that the gas bladder enables a fish to rise or sink in the water without having to swim.

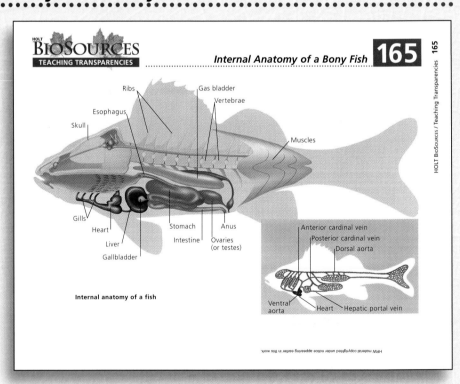

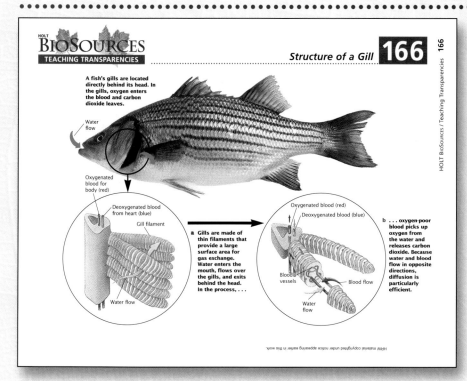

Teaching Strategies

- Use this transparency when discussing the structure and function of a fish's gills. Review the term *gill filament*.
- Point out that blood and water flow in opposite directions in the gill filament. Emphasize that the countercurrent arrangement of the gills enables oxygen to diffuse from water to blood more efficiently than if blood and water flowed in the same direction. A fish's gills can extract up to 85 percent of the available oxygen from water.
- Ask students to explain what happens to blood that has picked up oxygen in the gills. *(It flows to the rest of the body.)*

Evolution of Limbs 167

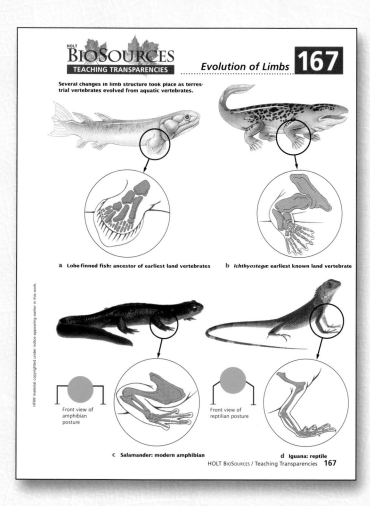

Teaching Strategies

- Use this transparency when discussing the evolution of limb structure.
- Have students study the illustration of the lobe-finned fish and compare the bones in its fins with the bones in the limbs of the *Ichthyostega,* the amphibian, and the reptile. Ask students to describe how these structures differ. *(The lobe-finned fish used its bones for swimming, but the other three organisms use theirs for walking.)*
- Inform students that *Ichthyostega* is an early amphibian that lived about 370 million years ago. Point out the strong hip, shoulder, and limb bones that enabled this animal to climb out of the water and walk around on land.

Teaching Strategies

- Use this transparency when discussing the major groups of bony fishes. Read each class name aloud so that students hear the correct pronunciation.
- Review the main characteristics of each group. Point out that the majority of the world's fishes are ray-finned species that belong to the class Osteichthyes.
- Ask students to name and describe the types of fish that are extinct.

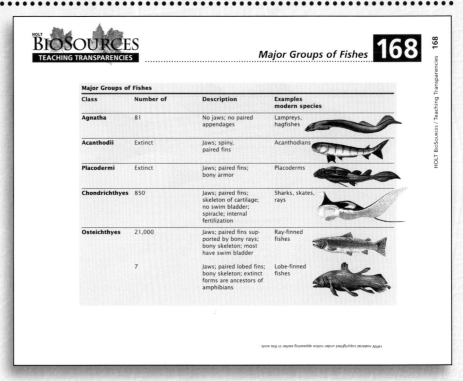

Major Groups of Fishes

Class	Number of	Description	Examples modern species
Agnatha	81	No jaws; no paired appendages	Lampreys, hagfishes
Acanthodii	Extinct	Jaws; spiny, paired fins	Acanthodians
Placodermi	Extinct	Jaws; paired fins; bony armor	Placoderms
Chondrichthyes	850	Jaws; paired fins; skeleton of cartilage; no swim bladder; spiracle; internal fertilization	Sharks, skates, rays
Osteichthyes	21,000	Jaws; paired fins supported by bony rays; bony skeleton; most have swim bladder	Ray-finned fishes
	7	Jaws; paired lobed fins; bony skeleton; extinct forms are ancestors of amphibians	Lobe-finned fishes

Teaching Strategies

- Use this transparency when describing the frog's life cycle. Review the terms *gills* and *metamorphosis*.
- Tell students that the eggs of frogs and toads are unprotected and jellylike. If the eggs are removed from water, they will dry out, and the embryos inside will die.
- Have students write a paragraph that describes metamorphosis in the frog life cycle.

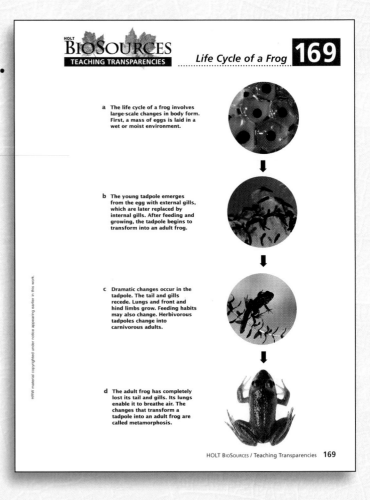

Life Cycle of a Frog

a The life cycle of a frog involves large-scale changes in body form. First, a mass of eggs is laid in a wet or moist environment.

b The young tadpole emerges from the egg with external gills, which are later replaced by internal gills. After feeding and growing, the tadpole begins to transform into an adult frog.

c Dramatic changes occur in the tadpole. The tail and gills recede. Lungs and front and hind limbs grow. Feeding habits may also change. Herbivorous tadpoles change into carnivorous adults.

d The adult frog has completely lost its tail and gills. Its lungs enable it to breathe air. The changes that transform a tadpole into an adult frog are called metamorphosis.

HOLT BioSources / Teaching Transparencies 169

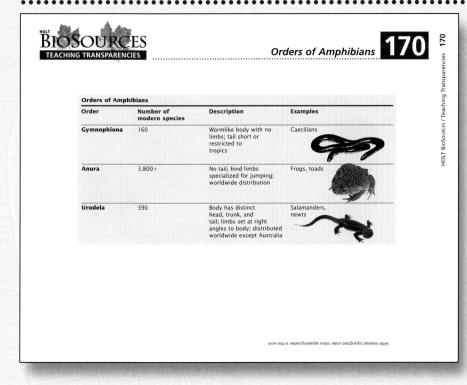

Teaching Strategies

- Use this transparency when discussing the characteristics of the three orders of amphibians.
- Point out that the majority of amphibians belong to the order Anura.
- Discuss the recent findings that show a decline in amphibian populations.

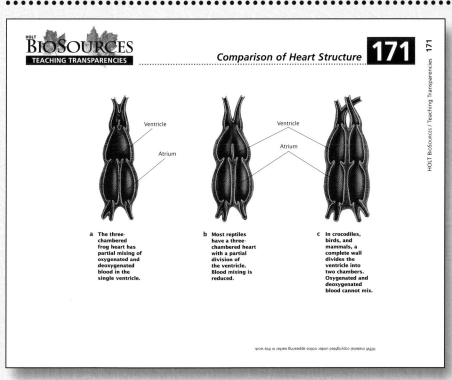

Teaching Strategies

- Use this transparency to explain the evolution of heart structure. Review the terms *atrium, ventricle,* and *septum.*
- Point out that the amphibian heart has two atria—one to receive deoxygenated blood from the body and the other to receive oxygenated blood that has traveled to the lungs for oxygenation.
- Point out that the ventricle in the reptilian heart is partially divided by a septum.
- Inform students that birds, crocodiles, and mammals have a ventricle that is divided into two completely different chambers.
- Point out the advantages or disadvantages of each heart's structure.

Teaching Strategies

- Use this transparency when comparing the structure of fish and amphibian circulation. Review the terms *gills* and *capillaries*.

- Use this diagram to contrast the heart of an amphibian with that of a fish. Point out the double-loop nature of the amphibian circulatory system. One loop carries oxygenated blood to the body and returns it to the heart. The other loop carries deoxygenated blood to the lungs, where the blood absorbs oxygen, then returns the blood to the heart to be pumped throughout the body.

- Have students describe the function of the pulmonary veins. *(They return oxygenated blood from the lungs to the heart.)*

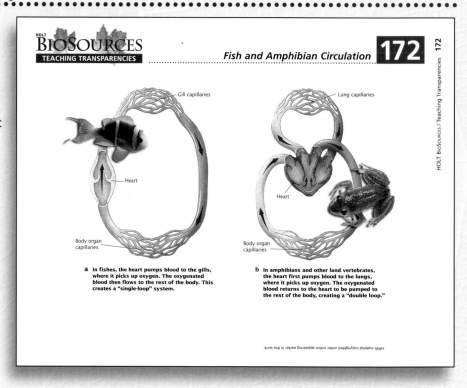

BIOSOURCES
TEACHING TRANSPARENCIES

Fish and Amphibian Circulation **172**

Gill capillaries

Lung capillaries

Heart

Heart

Body organ capillaries

Body organ capillaries

a In fishes, the heart pumps blood to the gills, where it picks up oxygen. The oxygenated blood then flows to the rest of the body. This creates a "single-loop" system.

b In amphibians and other land vertebrates, the heart first pumps blood to the lungs, where it picks up oxygen. The oxygenated blood returns to the heart to be pumped to the rest of the body, creating a "double loop."

Teaching Strategies

- Use this transparency when discussing the evolutionary relationship among the groups of living reptiles.

- Point out that turtles are thought to be most closely related to the early reptiles. Also point out the close relationship between lizards and snakes (snakes evolved from lizards) and between crocodiles and birds (both groups share a common ancestor).

- Have students identify which group of reptiles contains the immediate ancestors of birds *(dinosaurs)*.

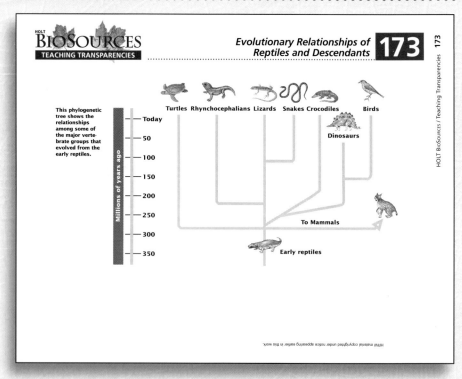

BIOSOURCES
TEACHING TRANSPARENCIES

Evolutionary Relationships of Reptiles and Descendants **173**

This phylogenetic tree shows the relationships among some of the major vertebrate groups that evolved from the early reptiles.

Turtles Rhynchocephalians Lizards Snakes Crocodiles Birds

Dinosaurs

Millions of years ago

Today
50
100
150
200
250
300
350

To Mammals

Early reptiles

Transparency 174

HOLT BioSources
TEACHING TRANSPARENCIES

Orders of Living Reptiles **174**

Order	Approximate Number of Living Species	Main Characteristics	Examples	
Squamata suborder Sauria	3,000	Lizards; limbs set at right angles to body; dry skin of scales; socketless teeth; heart with partially divided ventricle; most species are terrestrial, but a few are at least partially aquatic	Anoles, geckos, horned lizards	
Squamata suborder Serpentes	2,500	Snakes; no legs; scaly skin is shed periodically; socketless teeth; heart with partially divided ventricle; no external ear openings; most species are terrestrial, but a few are aquatic	Rattlesnakes, garter snakes	
Testudines	250	Body encased in shell of bony plates; sharp, horny jaw edges without teeth; vertebrae and ribs fused to shell; terrestrial and aquatic species	Turtles, tortoises, terrapins	
Crocodylia	22	Four-chambered heart; extended jaw with socketed teeth; five digits on forelimbs, four digits on hind limbs; live near or in water	Crocodiles, alligators	
Rhynchocephalia	2	Sole survivors of a group that largely disappeared about 100 million years ago. Skull like those of early Permian reptiles; fused, wedgelike, socketless teeth; terrestrial	Tuataras	

HFW material copyrighted under notice appearing earlier in this work.

HOLT BioSources / Teaching Transparencies **174**

Orders of Living Reptiles 174

Teaching Strategies

• Use this transparency when discussing the characteristics of the orders of living reptiles. Read the name of each order aloud so that students hear the correct pronunciation.

• Review the characteristics of each reptilian order. Point out that the majority of the world's reptiles belong to the order Squamata. Have students describe Squamata's two suborders and the reptiles they contain.

HOLT BioSources
TEACHING TRANSPARENCIES

Major Orders of Birds **175**

Order	Approximate Number of Living Species	Main Characteristics	Examples	
Passeriformes	5,276	Songbirds; perching feet; well-developed vocal organs; dependent young; largest bird order, containing 60 percent of all bird species	Sparrows, robins, warblers, crows, starlings, mockingbirds,	
Apodiformes	428	Small bodies, short legs, rapid wing beat; hummingbirds are the smallest birds	Hummingbirds, swifts	
Piciformes	383	Sharp, chisel-like bills for pounding through wood in search of insects; grasping feet	Woodpeckers, toucans, honeyguides	
Psittaciformes	340	Well-developed vocal organs; large powerful bills for crushing seeds	Parrots, cockatoos	
Charadriiformes	331	Shorebirds; typically with long, slender, probing bills and long, stiltlike legs	Gulls, terns, plovers, auks, sandpipers	
Columbiformes	303	Stout bodies; perching feet	Pigeons, doves	
Falconiformes	288	Birds of prey; day-active carnivores; sharp pointed beaks for tearing flesh; keen vision; strong fliers	Eagles, hawks, falcons, vultures	
Galliformes	268	Rounded bodies; often limited flying ability	Chickens, quail, grouse, pheasants	

HFW material copyrighted under notice appearing earlier in this work.

HOLT BioSources / Teaching Transparencies **175**

Major Orders of Birds 175

Teaching Strategies

• Use this transparency when describing the eight largest orders of birds. Read the name of each order aloud so that students hear the correct pronunciation.

• Review the main characteristics of each order. Point out that most of the common birds in North America belong to the order Passiformes. Passiformes is the largest order of terrestrial vertebrates and includes sparrows, robins, crows, warblers, finches, and starlings.

• Have students name the order to which chickens belong *(Galliformes)*.

Teaching Strategies

- Use this transparency when discussing the eight smallest orders of birds. Read the name of each order aloud so that students hear the correct pronunciation.
- Review the main characteristics of each order. Have students compare the order Strigiformes (owls) with the order Falconiformes (hawks). *(The members of both orders are carnivores and have sharp, curved beaks for tearing flesh and sharp talons for seizing prey. Owls feed at night and have very large eyes and sensitive hearing. Hawks are active during the day and have very keen vision.)*
- Have students name one swimming adaptation of the order Anseriformes *(webbed feet).*

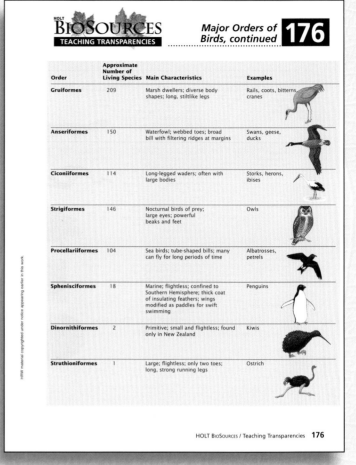

HOLT BioSources
TEACHING TRANSPARENCIES

Major Orders of Birds, continued **176**

Order	Approximate Number of Living Species	Main Characteristics	Examples
Gruiformes	209	Marsh dwellers; diverse body shapes; long, stiltlike legs	Rails, coots, bitterns, cranes
Anseriformes	150	Waterfowl; webbed toes; broad bill with filtering ridges at margins	Swans, geese, ducks
Ciconiiformes	114	Long-legged waders; often with large bodies	Storks, herons, ibises
Strigiformes	146	Nocturnal birds of prey; large eyes; powerful beaks and feet	Owls
Procellariiformes	104	Sea birds; tube-shaped bills; many can fly for long periods of time	Albatrosses, petrels
Sphenisciformes	18	Marine; flightless; confined to Southern Hemisphere; thick coat of insulating feathers; wings modified as paddles for swift swimming	Penguins
Dinornithiformes	2	Primitive; small and flightless; found only in New Zealand	Kiwis
Struthioniformes	1	Large; flightless; only two toes; long, strong running legs	Ostrich

HRW material copyrighted under notice appearing earlier in this work.

HOLT BioSources / Teaching Transparencies **176**

Teaching Strategies

- Use this transparency when discussing the characteristics of monotremes, marsupials, and the five largest orders of placental mammals. Review the terms *monotreme, marsupial*, and *placental*.
- Review the main characteristics of the mammals listed in the table. Have students design a graphic organizer that illustrates the similarities and differences among monotremes, marsupials, and placental mammals.
- Point out that the order Rodentia is the largest mammalian order, containing about 40 percent of mammalian species. Most of the members of this order are small and reproduce rapidly. The identifying feature of rodents is the presence of four continuously growing incisors in each jaw.

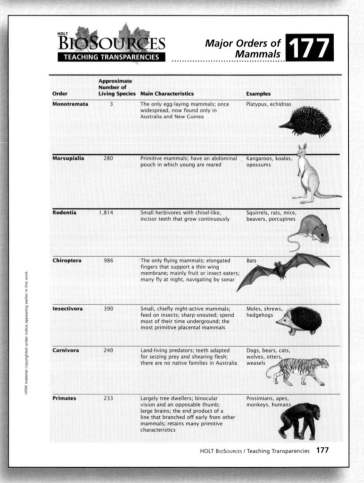

HOLT BioSources
TEACHING TRANSPARENCIES

Major Orders of Mammals **177**

Order	Approximate Number of Living Species	Main Characteristics	Examples
Monotremata	3	The only egg-laying mammals; once widespread, now found only in Australia and New Guinea	Platypus, echidnas
Marsupialia	280	Primitive mammals; have an abdominal pouch in which young are reared	Kangaroos, koalas, opossums
Rodentia	1,814	Small herbivores with chisel-like, incisor teeth that grow continuously	Squirrels, rats, mice, beavers, porcupines
Chiroptera	986	The only flying mammals; elongated fingers that support a thin wing membrane; mainly fruit or insect eaters; many fly at night, navigating by sonar	Bats
Insectivora	390	Small, chiefly night-active mammals; feed on insects; sharp-snouted; spend most of their time underground; the most primitive placental mammals	Moles, shrews, hedgehogs
Carnivora	240	Land-living predators; teeth adapted for seizing prey and shearing flesh; there are no native families in Australia	Dogs, bears, cats, wolves, otters, weasels
Primates	233	Largely tree dwellers; binocular vision and an opposable thumb; large brains; the end product of a line that branched off early from other mammals; retains many primitive characteristics	Prosimians, apes, monkeys, humans

HRW material copyrighted under notice appearing earlier in this work.

HOLT BioSources / Teaching Transparencies **177**

BIOSOURCES
TEACHING TRANSPARENCIES
Major Orders of
Mammals, continued **178**

Order	Approximate Number of Living Species	Main Characteristics	Examples
Artiodactyla	211	Hoofed mammals with two or four toes; large herbivores; most are grass eaters	Sheep, pigs, cattle, deer, giraffes
Cetacea	79	Aquatic, streamlined bodies; front limbs modified into broad flippers; no hind limbs; nostrils are blowholes on top of head; hairless except on muzzle	Whales, dolphins, porpoises
Lagomorpha	69	Rodentlike mammals with four upper incisors, rather than the two seen in rodents; hind legs often longer than forelegs, an adaptation for jumping	Rabbits, hares, pikas
Pinnipedia	34	Marine carnivores with limbs modified for swimming; feed mainly on fish	Seals, sea lions, walruses
Edentata	30	Mostly insect eaters; many are toothless, but some have degenerate, peglike teeth	Sloths, anteaters, armadillos
Perissodactyla	17	Hoofed mammals with one or three toes; herbivores with teeth adapted for chewing	Horses, zebras, rhinoceroses, tapirs
Proboscidea	2	Enormous herbivores with long trunks; two upper incisors elongated as tusks; the largest living land animals	Elephants

Teaching Strategies

- Use this transparency when discussing the characteristics of the seven smallest orders of mammals.
- Describe the differences between pinnipeds and cetaceans. Both groups have streamlined bodies and fins. However, pinnipeds spend time on land, which cetaceans cannot do.
- Have students contrast the order Lagomorpha with the order Rodentia.
- Ask students to name the order to which armadillos belong *(Edentata)*.

BIOSOURCES
TEACHING TRANSPARENCIES
Orders of
Extinct Reptiles **179**

Order	Approximate Number of Living Species	Main Characteristics	Examples
Ornithischia	Extinct	Mostly plant-eating dinosaurs with two pelvic bones facing backwards; hole in the skull in front of eye socket; legs positioned beneath the body; over 150 genera	Triceratops, Stegosaurus, Iguanodon
			Stegosaurus
Saurischia	Extinct	Flesh-eating and plant-eating dinosaurs with one pelvic bone facing forward, the other backward; terrestrial with three or five toes; hole in skull in front of eye socket; legs positioned beneath body; over 200 genera	Tyrannosaurus, Brontosaurus, Brachiosaurus
			Tyrannosaurus rex
Pterosauria	Extinct	Flying reptiles with wings of skin between fourth finger and body; wing span of early (Jurassic) *Rhamphorhynchus* was typically 60 cm (2 ft.), of later (Cretaceous) *Pteranodon* over 7.5 m (25 ft.)	Pteranodon, Pterodactylus, Rhamphorynchus
			Rhamphorynchus
Plesiosauria	Extinct	Marine reptiles with very large paddle-shaped fins, a barrel shaped body, and long jaws with sharp teeth; some had a snakelike neck twice as long as the body; others had a short neck and elongated skull about 3.7 m (12 ft.) in length	Plesiosaurus, Elasmosaurus, Kronosaurus
			Plesiosaurus
Ichthyosauria	Extinct	Marine reptiles with streamlined bodies up to 3 m (10 ft.) in length; the four legs modified into balancing fins; apparently fast swimmers, with many body similarities to modern fishes such as tuna or mackerel	Ichthyosaurus

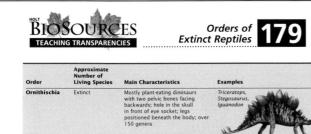

Orders of
Extinct Reptiles **179**

Teaching Strategies

- Use this transparency when discussing the orders of extinct reptiles. Read the name of each order aloud so that students hear the correct pronunciation.
- Have students relate the characteristics of each order to the structure of the representative organism in the illustration.
- Have students name examples of the species that are classified in the order Saurischia *(Tyrannosaurus, Brachiosaurus,* and *Apatosaurus)*.

Teaching Strategies

- Use this transparency when discussing the evolution of lung structure.
- Point out that amphibians have simple lungs with a relatively small internal surface area. Explain that amphibians cannot absorb oxygen as rapidly through their lungs as can reptiles, mammals, and birds.
- Have students explain why amphibians do not have countercurrent flow. *(Because air flows into and out of the lungs, it cannot flow in the opposite direction of blood in the lung capillaries.)*

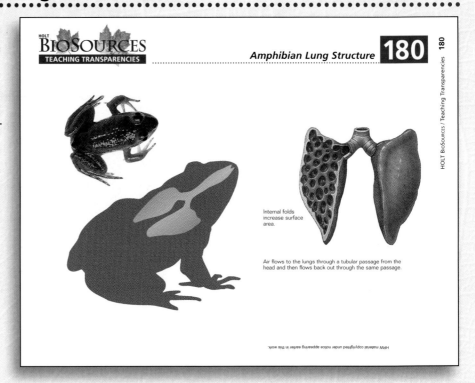

Internal folds increase surface area.

Air flows to the lungs through a tubular passage from the head and then flows back out through the same passage.

HRW material copyrighted under notice appearing earlier in this work.

181 *Mammalian Lung Structure*

Teaching Strategies

- Use this transparency when discussing the evolution of lung structure.
- Inform students that mammals have more alveoli than do reptiles. As a result, mammals can absorb oxygen more rapidly. Students should be able to predict that mammals have higher metabolic rates than reptiles. However, point out that countercurrent flow does not occur in the lungs of mammals.
- Tell students that human lungs have an internal surface area of 80 m². Ask them to explain how such a large area can fit into the chest. *(The lungs contain many alveoli, which constitute this large area.)*

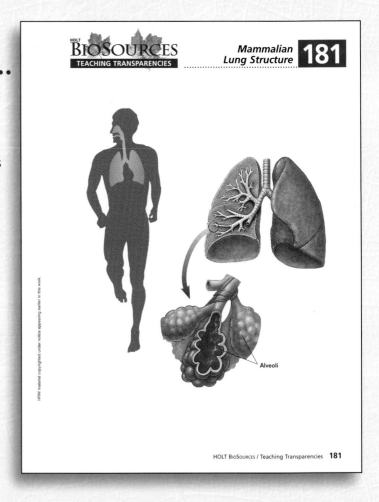

Alveoli

HRW material copyrighted under notice appearing earlier in this work.

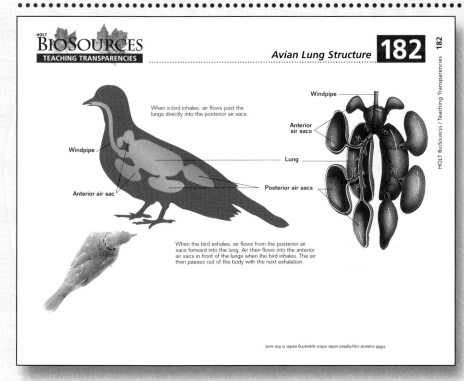

Avian Lung Structure **182**

When a bird inhales, air flows past the lungs directly into the posterior air sacs.

Windpipe

Windpipe

Anterior air sacs

Anterior air sac

Lung

Posterior air sacs

When the bird exhales, air flows from the posterior air sacs forward into the lung. Air then flows into the anterior air sacs in front of the lungs when the bird inhales. The air then passes out of the body with the next exhalation.

Teaching Strategies

• Use this transparency when discussing the evolution of lung structure.

• Inform students that birds have a high demand for oxygen. Birds meet this demand by having lungs with one-way air flow. Point out that oxygen-rich air is always in the lungs. Also, blood and air flow cross each other in an arrangement called cross-current flow, which is more efficient than the two-way flow of mammalian lungs.

• Have students trace the path a volume of air follows through the respiratory system of a bird *(windpipe, posterior air sac, lung, anterior air sac, windpipe).*

Comparison of Vertebrate Excretory Systems 183

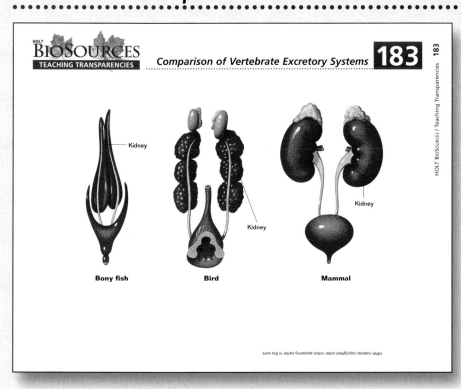

Comparison of Vertebrate Excretory Systems **183**

Kidney

Kidney

Kidney

Kidney

Bony fish   **Bird**   **Mammal**

Teaching Strategies

• Use this transparency when comparing the excretory systems of fishes, birds, and mammals.

• Remind students that terrestrial animals tend to lose body water through evaporation. Thus, an excretory system that reduces the amount of water eliminated in the urine is advantageous. Point out that the excretory systems of birds and mammals produce concentrated urine.

• Have students explain why it is advantageous for a bird to have no urinary bladder to store urine. *(Urine contains water and is therefore heavy.)*

Teaching Strategies

- Use this transparency when discussing the anatomy of the knee joint. Review the terms *cartilage, ligament, tibia,* and *fibula.*

- Point out that bones of the knee joint do not touch. They are tipped by cartilage and separated by a fluid-filled sac. This sac surrounds the joint and contains lubricating fluid. Inform students that prolonged kneeling can keep the fluid from circulating through the joint, resulting in "creaky" knees often experienced when one rises from the kneeling position.

- Have students describe the function of the ligament shown in the figure. *(It connects the femur to the tibia.)*

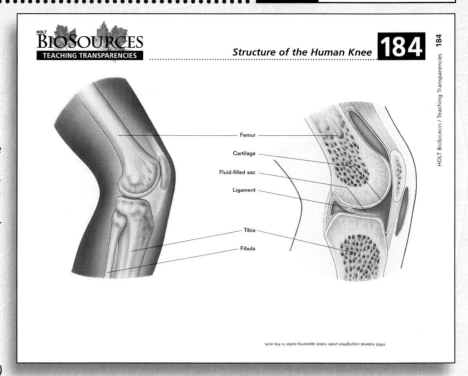

BIOSOURCES
TEACHING TRANSPARENCIES

Structure of the Human Knee **184**

HOLT BioSources / Teaching Transparencies   184

- Femur
- Cartilage
- Fluid-filled sac
- Ligament
- Tibia
- Fibula

HRW material copyrighted under notice appearing earlier in this work.

Teaching Strategies

- Use this transparency to illustrate the three basic types of joints in the human body: immovable joints, slightly movable joints, and freely movable joints.

- Ask students to identify which kinds of joints are cushioned by pads of cartilage *(slightly movable joints and freely movable joints).*

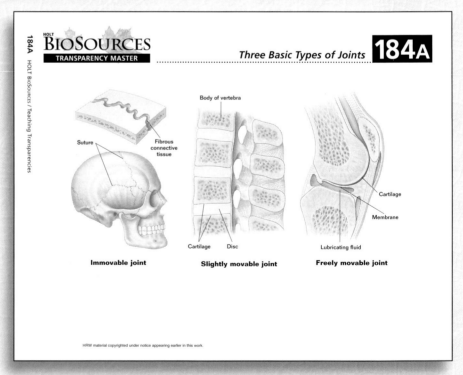

184A   HOLT BioSources / Teaching Transparencies

BIOSOURCES
TRANSPARENCY MASTER

Three Basic Types of Joints **184A**

Body of vertebra

Suture

Fibrous connective tissue

Cartilage

Cartilage   Disc

Membrane

Lubricating fluid

Immovable joint   **Slightly movable joint**   **Freely movable joint**

HRW material copyrighted under notice appearing earlier in this work.

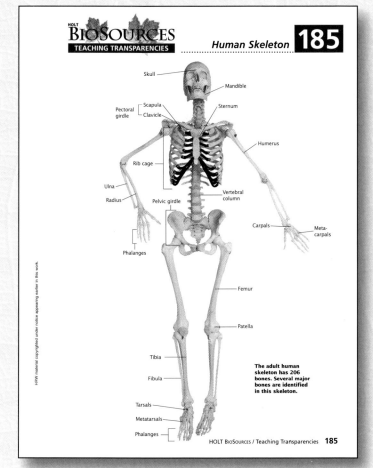

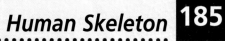

Teaching Strategies

- Use this transparency when discussing the major bones that make up the human skeleton. Inform students that all vertebrate skeletons are customarily divided into two components: the axial skeleton and the appendicular skeleton. The axial skeleton consists of the skull, vertebral column, and ribs. All the other bones belong to the appendicular skeleton.
- Review the major human skeletal bones labeled in the figure.
- Point out the following types of joints: ball-and-socket, pivot, plane, hinge, and suture. Have students describe the movement each joint allows.

Electrocardiogram 185A

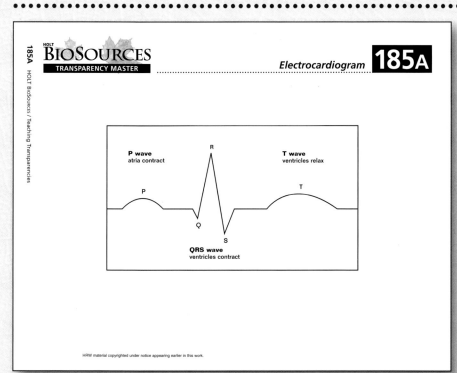

Teaching Strategies

- Use this transparency to illustrate the results of an electrocardiogram.
- Remind students that the P, QRS, and T waves shown in the figure occur each time the heart beats. Tell them that cardiologists, physicians who specialize in the heart, are trained to interpret electrocardiograms.
- Have students explain what an electrocardiogram measures. *(It measures the tiny electrical impulses produced by the heart muscle when it contracts and relaxes.)*

Teaching Strategies

- Use this transparency to illustrate the structure of bone tissue. Review the terms *spongy bone, compact bone,* and *marrow.*

- Have students describe the appearance of both compact bone and spongy bone. *(Compact bone is made up of concentric rings of bone deposited around microscopic channels called Haversian canals. Each cylindrical canal is a narrow tunnel through which blood vessels and nerves pass. Spongy bone is made up of hardened fibers interlaced with many spaces.)*

Where is marrow found and what is its function?

(Marrow fills the entire center of spongy bone and forms blood cells. In older bones, marrow stores fat.)

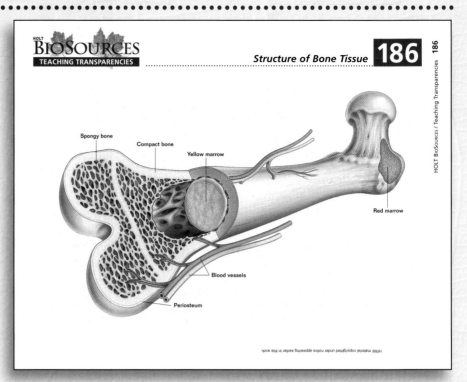

BioSources
TEACHING TRANSPARENCIES

Structure of Bone Tissue **186**

- Spongy bone
- Compact bone
- Yellow marrow
- Red marrow
- Blood vessels
- Periosteum

HOLT BioSources / Teaching Transparencies 186

Teaching Strategies

- Use this transparency to illustrate the structure of the human lymphatic system. Review the term *lymphatic node.*

- Point out the major features of the lymphatic system, including the lymphatic vessels and the lymph nodes. Lymph nodes filter foreign substances from the lymph that travels through them.

- Have students explain how the lymphatic system helps your body maintain a constant blood volume. *(The lymphatic system collects fluid that leaks from the blood and returns it to the cardiovascular system.)*

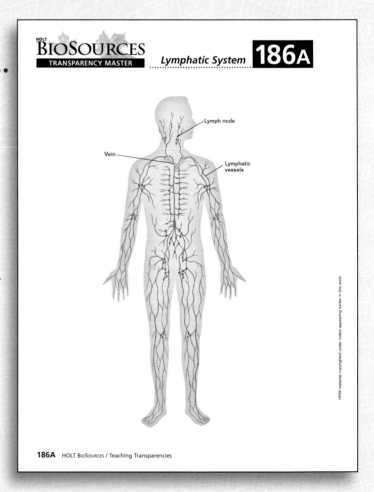

BioSources
TRANSPARENCY MASTER

Lymphatic System **186A**

- Lymph node
- Vein
- Lymphatic vessels

186A HOLT BioSources / Teaching Transparencies

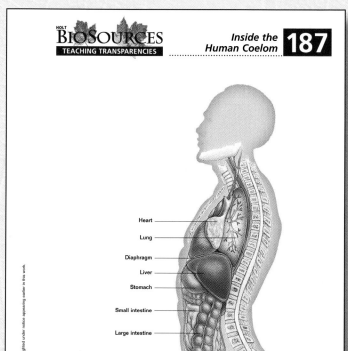

HOLT
BIOSOURCES
TEACHING TRANSPARENCIES

Inside the
Human Coelom 187

Heart
Lung
Diaphragm
Liver
Stomach
Small intestine
Large intestine

HOLT BioSources / Teaching Transparencies **187**

Teaching Strategies

- Use this transparency to illustrate the two compartments in the human coelom. Point out the diaphragm, the sheet of muscle that aids in respiration. Have students identify the organs located above the diaphragm *(lungs and heart)* and below the diaphragm *(organs of the digestive system)*.

- Review the evolutionary and developmental history of the coelom. Remind students that during development, the coelom forms within the mesoderm. The first coelom appeared in mollusks.

How does food pass through the thoracic cavity into the stomach, which is in the abdominal cavity?

(Food passes through the esophagus.)

Composition of Blood 187A

HOLT
BIOSOURCES
TRANSPARENCY MASTER

Composition of Blood **187A**

Components of Blood	Function
Plasma portion (60% of total blood volume)	
Water	Acts as solvent
Metabolites and wastes, salts and ions, proteins	Play diverse roles (nourish cells, catalyze chemical reactions, act as chemical messengers, maintain blood volume, fight infection, etc.)
Cellular portion (40% of total blood volume)	
Red blood cells	$O_2 + CO_2$ transport
White blood cells	Produce antibodies, ingest foreign materials
Platelets	Aid in clotting blood

Teaching Strategies

- Use this transparency when discussing the components of blood. Review the terms *plasma* and *platelet*.

- Emphasize the complex composition and diverse functions of blood by reading this table aloud to students. Remind them that blood is a type of connective tissue.

- Have students explain why red blood cells must be small in size. *(If they were much larger, they could not fit into capillaries, and tissues would never receive the oxygen they need.)*

Teaching Strategies

- Remind students that the skin is the largest organ of the body. Emphasize that skin is a very complex organ even though this diagram shows a simplified view of its structure.

- Review the structures found in each layer of skin. Be sure that students locate hair follicles and shafts, oil glands, sweat glands, and sweat pores. Emphasize which structures pass through both layers of skin (*sweat pores and hair shafts*).

What is the thickest layer of human skin? *(the dermal layer)*

Which layer of skin is most important to the sense of touch? *(the dermal layer)*

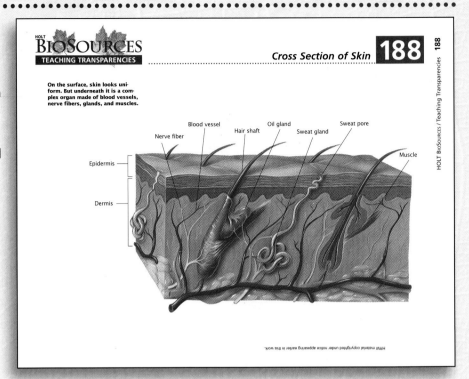

HOLT BIOSOURCES
TEACHING TRANSPARENCIES

Cross Section of Skin **188**

On the surface, skin looks uniform. But underneath it is a complex organ made of blood vessels, nerve fibers, glands, and muscles.

Nerve fiber — Blood vessel — Hair shaft — Oil gland — Sweat gland — Sweat pore — Muscle

Epidermis

Dermis

Teaching Strategies

- Use this transparency to illustrate the blood-clotting process.

- Trace the blood-clotting process depicted in this figure. Tell students that this is a simplified version of a complex, multistep pathway involving as many as 12 chemical factors either found in plasma or released from megakaryocytes or damaged tissues. The chemical factors include calcium ions, lipoprotein complexes, and protein components. Emphasize that the fibrin net is the end result of the blood-clotting process.

- Have students name the chemical composition of fibrin (*protein*).

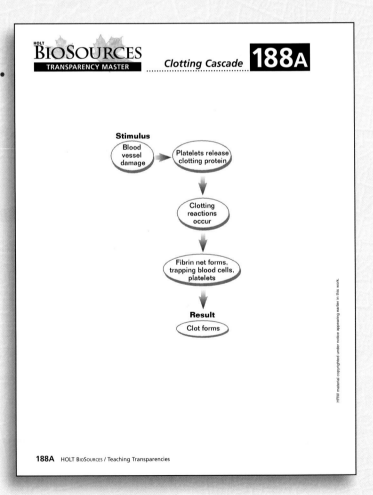

HOLT BIOSOURCES
TRANSPARENCY MASTER

Clotting Cascade **188A**

Stimulus

Blood vessel damage → Platelets release clotting protein

↓

Clotting reactions occur

↓

Fibrin net forms, trapping blood cells, platelets

↓

Result

Clot forms

188A HOLT BIOSOURCES / Teaching Transparencies

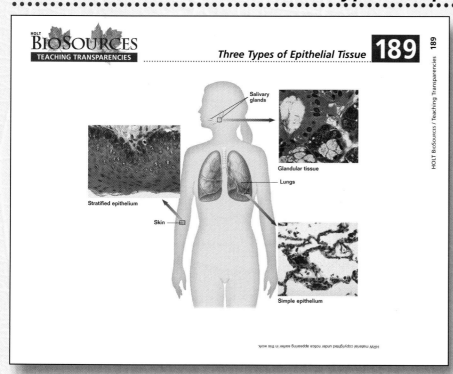

HOLT **BioSources**
TEACHING TRANSPARENCIES

Three Types of Epithelial Tissue 189

189

HOLT BioSources / Teaching Transparencies

Salivary glands

Glandular tissue

Lungs

Stratified epithelium

Skin

Simple epithelium

HRW material copyrighted under notice appearing earlier in this work.

Teaching Strategies

- Use this transparency when discussing the differences in structure and location among the three types of epithelial tissues. Review the terms *gland* and *epithelium*.
- Inform students that the epithelial layer lining the lungs is only one cell layer thick. Thus, oxygen must travel only a short distance to reach the blood.
- Ask students why the skin is not composed of epithelial tissue that is only one layer thick. (*The skin must be thick so that it can resist abrasion and wear. Two of its many functions are to keep pathogens out and the body's fluids in.*)

189A

HOLT BioSources / Teaching Transparencies

HOLT **BioSources**
TRANSPARENCY MASTER

Blood Types 189A

Type	Antigen on the RBC	Antibodies in Plasma	Can Receive Blood From	Can Donate Blood From
A	A	B	O, A	A, AB
B	B	A	O, B	B, AB
AB	A, B	None	O, A, B, AB	AB
O	None	A, B	O	O, A, B, AB

HRW material copyrighted under notice appearing earlier in this work.

Teaching Strategies

- Use this transparency to summarize the four possible blood types. Review the terms *antigen* and *antibody*.
- Review this table with students to reinforce the material presented in the text. Stress that antigens are found on the surface of the erythrocytes (RBCs), while antibodies are dissolved in plasma.
- Have students name an important antigen, other than the A or B antigen, that is found on the surface of the red blood cells (*Rh factor*).

Teaching Strategies

- Use this transparency when discussing the structures that make up skeletal muscle. Review the terms *microfibril, sarcomere, Z line, actin,* and *myosin.*

- Point out the levels of organization shown in this figure. Help students understand that muscles are composed of many bundles of muscle fibers, that a muscle fiber contains many myofibrils, and that a myofibril consists of actin and myosin filaments.

- Have students explain the mechanism of muscle contraction. *(A muscle contraction begins at the molecular level of organization, when actin and myosin filaments slide past one another and cause the muscle to shorten.)*

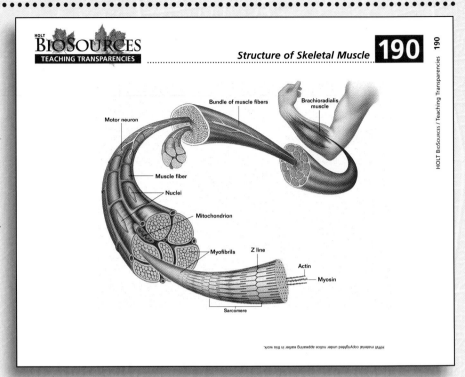

Teaching Strategies

- Read this nutrition label aloud to students. Point out the segments of it containing values for the fats contained in the food. Discuss the difference between saturated, monounsaturated, and saturated fats.

- Emphasize that naturally saturated animal fats as well as hydrogenated vegetable oils are considered saturated fats even though the label may not distinguish them as such.

- Ask students to describe the health risks associated with eating a diet that is too high in saturated fat *(heart disease, stroke, certain types of cancer, obesity).* Encourage students to be more aware of the foods they eat.

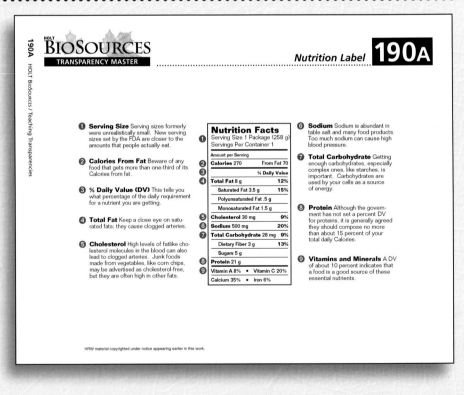

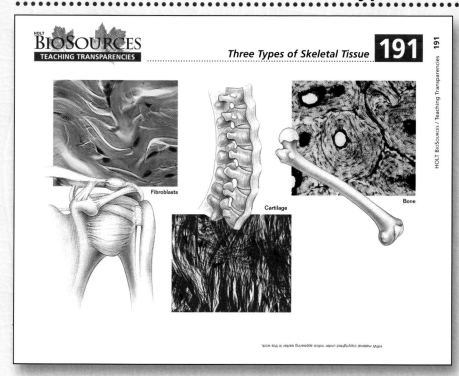

BioSources
TEACHING TRANSPARENCIES

Three Types of Skeletal Tissue 191

HOLT BioSources / Teaching Transparencies 191

Fibroblasts

Cartilage

Bone

HRW material copyrighted under notice appearing earlier in this work.

Teaching Strategies

- Use this transparency to contrast the microscopic structures of the three different kinds of tissues in the skeletal system. Review the terms *fibroblast* and *cartilage*.

- Point out that cartilage is a connective tissue in which the intercellular matrix consists of very thin proteinaceous fibers. It is firm, but not as hard or brittle as bone.

- Have students design a graphic organizer that illustrates the structure and function of the three types of skeletal tissues shown in this figure.

BioSources
TRANSPARENCY MASTER

Vitamins 191A

Vitamin	Food Sources	Role	
Water-Soluble			
Vitamin B₁ (thiamin)	Most vegetables, nuts, organ meats	Carbohydrate metabolism, helps nerves and heart to function properly	Digestive disturbances, impaired senses
Vitamin B₂ (riboflavin)	Fish, poultry, cheese, yeast, green vegetables	Needed for healthy skin and tissue repair, carbohydrate metabolism	Blurred vision, cataracts, cracking of skin, lesions of intestinal lining
Vitamin B₃ (niacin)	Whole grains, fish, poultry, liver, tomatoes, legumes, potatoes	Keeps skin healthy, carbohydrate metabolism	Mental disorders, diarrhea, inflamed skin
Vitamin B₁₂ (cobalamin)	Meat, poultry, green vegetables, milk, dairy products	Needed for formation of red blood cells	Reduced number of red blood cells
Vitamin C (ascorbic acid)	Citrus fruits, strawberries, potatoes	Needed for wound healing, healthy gums and teeth	Swollen bleeding gums, loose teeth, and slow-healing wounds
Fat-Soluble			
Vitamin A (retinol)	Carrots, green leafy vegetables, butter, eggs, liver, sweet potatoes	Keeps eyes and skin healthy, needed for strong bones and teeth	Infections of urinary and digestive systems, night blindness
Vitamin D (cholecalciferol)	Salmon, tuna, fish liver oils, fortified milk	Calcium and phosphorus metabolism, needed for strong bones and teeth	Bone deformities in children, loss of muscle tone
Vitamin E (tocopherol)	Many foods, especially wheat germ oil and olives	Protects cell membranes from damage by reactive oxygen compounds	Reduced number of red blood cells; nerve tissue damage in infants
Vitamin K	Leafy green vegetables, liver, cauliflower	Necessary for normal blood clotting	Bleeding caused by a prolonged clotting time

HRW material copyrighted under notice appearing earlier in this work.

Vitamins 191A

Teaching Strategies

- Remind students that vitamins are required for good health. Tell them that most vitamins function as parts of enzymes or coenzymes. Enzymes and coenzymes control the chemical reactions that release energy in the cells of their bodies.

- Ask students to explain why vitamins are required in only small amounts. *(Enzyme and coenzyme molecules inside cells are used over and over again.)*

Teaching Strategies

• Use this transparency when describing the events underlying muscle contraction: attachment of myosin heads to actin, bending of myosin heads, release of bonds between myosin and actin, and recocking of myosin heads for the next cycle.

• Emphasize that when a muscle cell contracts, it shortens as its sarcomeres and the distance between Z lines decrease.

• Have students describe a Z line *(proteins at opposite ends of a sarcomere that are attachment points for actin).*

Is myosin attached to a Z line?
(no)

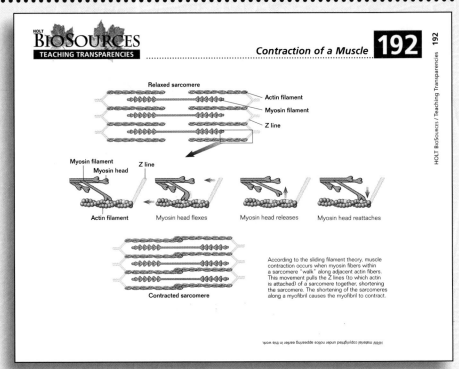

BIOSOURCES
TEACHING TRANSPARENCIES

Contraction of a Muscle **192**

Relaxed sarcomere
- Actin filament
- Myosin filament
- Z line

Myosin filament
Myosin head
Z line

Actin filament | Myosin head flexes | Myosin head releases | Myosin head reattaches

Contracted sarcomere

According to the sliding filament theory, muscle contraction occurs when myosin fibers within a sarcomere "walk" along adjacent actin fibers. This movement pulls the Z lines (to which actin is attached) of a sarcomere together, shortening the sarcomere. The shortening of the sarcomeres along a myofibril causes the myofibril to contract.

Teaching Strategies

• Remind students that although trace elements are required in only very minute amounts, each one has an important function. Tell students to imagine a person who is deficient in all the trace elements listed in the table. He or she would most likely be in very ill health.

• Have students name two elements that are not trace elements but that are required by your body *(calcium and sodium).*

BIOSOURCES
TRANSPARENCY MASTER

Trace Elements **192A**

	Best Sources	Role
Iodine	Iodized salt, seafood, plants grown in high-iodine soil	Synthesis of thyroid hormone
Cobalt	Leafy vegetables, liver, kidney	Synthesis of vitamin B_{12}
Zinc	Meat, shellfish, dairy products	Synthesis of digestive enzymes
Molybdenum	Legumes, cereals, milk	Protein synthesis
Manganese	Whole grains, nuts, legumes	Hemoglobin synthesis, urea formation
Selenium	Meat, seafood, cereal grains	Preventing chromosome breakage

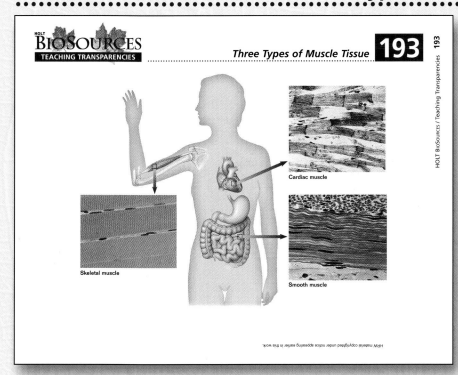

HOLT
BIOSOURCES
TEACHING TRANSPARENCIES

Three Types of Muscle Tissue **193**
HOLT BioSources / Teaching Transparencies 193

Cardiac muscle

Skeletal muscle

Smooth muscle

Teaching Strategies

- Use this transparency when discussing the function and structure of the three types of muscle tissues: skeletal, cardiac, and smooth muscle. Review the characteristics of each tissue type.

- Point out the latticework structure of cardiac muscle. Emphasize that the cells in cardiac muscle are electronically connected, so that impulses travel from cell to cell, causing a wave of contraction. Also point out that smooth muscles are not under voluntary control but that skeletal muscles are.

- Ask students if cardiac muscle is under conscious control. *(No, but heart rate is influenced by emotional state.)*

Digesting Food Molecules **193A**

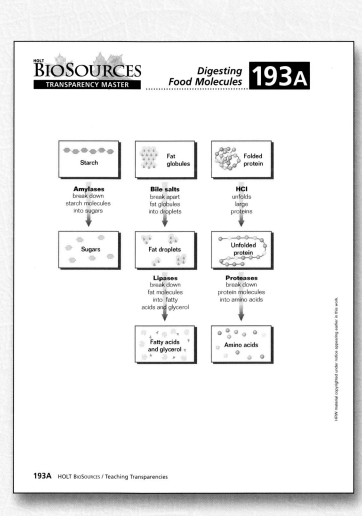

HOLT
BIOSOURCES
TRANSPARENCY MASTER

Digesting Food Molecules **193A**

Starch

Fat globules

Folded protein

Amylases
break down starch molecules into sugars

Bile salts
break apart fat globules into droplets

HCl
unfolds large proteins

Sugars

Fat droplets

Unfolded protein

Lipases
break down fat molecules into fatty acids and glycerol

Proteases
break down protein molecules into amino acids

Fatty acids and glycerol

Amino acids

Teaching Strategies

- Remind students that during digestion, complex food molecules are dismantled into their smaller components, like sugars, amino acids, glycerols, and fatty acids. Only these small molecules can pass through cell membranes.

- Have students classify the molecules amylase, lipase, and protease. *(They are enzymes.)*

Teaching Strategies

- Use this transparency when discussing the basic structure and function of a neuron. Review the terms *dendrite, axon, synapse,* and *neurotransmitter.*

- Trace the path an impulse travels along a neuron, beginning with the short fingerlike dendrites. Point out that the axon is covered with an intermittent myelin sheath and that the impulse skips from node to node. When the impulse reaches the axon's tip, neurotransmitters are released into the synapse and carry the signal to the adjacent cell.

- Have students write a paragraph that describes a nerve impulse traveling down their hand as they write.

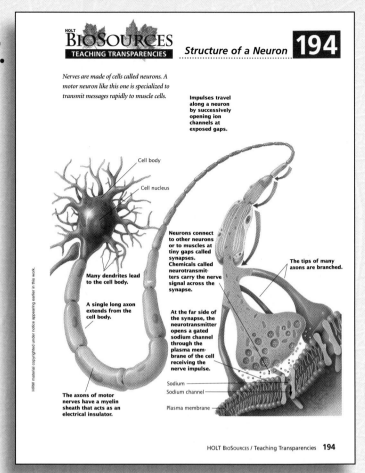

HOLT BioSources / TEACHING TRANSPARENCIES

Structure of a Neuron 194

Nerves are made of cells called neurons. A motor neuron like this one is specialized to transmit messages rapidly to muscle cells.

Impulses travel along a neuron by successively opening ion channels at exposed gaps.

Cell body

Cell nucleus

Many dendrites lead to the cell body.

A single long axon extends from the cell body.

Neurons connect to other neurons or to muscles at tiny gaps called synapses. Chemicals called neurotransmitters carry the nerve signal across the synapse.

The tips of many axons are branched.

At the far side of the synapse, the neurotransmitter opens a gated sodium channel through the plasma membrane of the cell receiving the nerve impulse.

The axons of motor nerves have a myelin sheath that acts as an electrical insulator.

Sodium

Sodium channel

Plasma membrane

HOLT BioSources / Teaching Transparencies **194**

Teaching Strategies

- Use this table to review the human body's major glands and the hormones they produce.

- Ask students to identify the target organs for each hormone and the effects each hormone has on its target organ(s).

- Have students refer to the table to answer the following questions:

Which hormones act on the kidneys?
(ADH, cortisol, and parathyroid hormone)

Which hormones act on all tissues?
(growth hormone, aldosterone, insulin, and thyroxine)

Which hormones stimulate the production of other hormones?
(ACTH, FSH, LH, and TSH)

HOLT BioSources / TRANSPARENCY MASTER

Major Endocrine Glands and Hormones 194A

Gland	Hormone	Target Tissue	Effects
Pituitary gland	**Anterior lobe**		
	Adrenocorticotropic hormone (ACTH)	Adrenal glands	Stimulates the production of steroid hormones
	Follicle-stimulating hormone (FSH)	Ovaries and testes	Regulates the development of male and female gametes; stimulates the production of testosterone (male sex hormone) in males
	Luteinizing hormone (LH)	Ovaries and testes	Stimulates the release of an egg (ovulation) from an ovary; stimulates testosterone production by the testes
	Prolactin	Mammary glands	Stimulates milk production in breasts
	Somatotropin, or growth hormone (GH)	All tissues	Promotes protein synthesis; stimulates growth of muscles and bones
	Thyroid-stimulating hormone (TSH)	Thyroid gland	Stimulates production of thyroxin by the thyroid gland
	Posterior lobe		
	Antidiuretic hormone (ADH)	Kidneys, blood vessels	Stimulates reabsorption of water; constricts blood vessels
	Oxytocin	Mammary glands, uterus	Stimulates uterine contractions and milk secretion
Adrenal glands	**Cortex**		
	Aldosterone	All tissues	Controls salt (sodium and potassium) and water balance
	Cortisol	Kidneys	Stimulates metabolism of carbohydrates, lipids, and proteins; raises blood sugar
	Medulla		
	Epinephrine (adrenaline) and norepinephrine	Skeletal and cardiac muscle, blood vessels	Initiates the response to stress; increases metabolic rate, heart rate, and blood pressure; dilates blood vessels; raises blood sugar
Islets of Langerhans (Pancreas)	Glucagon	Liver, fatty tissues	Stimulates conversion of glycogen to glucose; raises blood sugar
	Insulin	All tissues	Stimulates conversion of glucose to glycogen; lowers blood sugar

Anterior lobe

Posterior lobe

Cortex

Medulla

194A HOLT BioSources / Teaching Transparencies

BIOSOURCES
TEACHING TRANSPARENCIES

Physiology of a Nerve Impulse **195**

Resting potential

Inside cell

[Na⁺] [K⁺]

Outside cell

[Na⁺] [K⁺]

Axon

Path of impulse

Action potential

Na⁺

K⁺

Na⁺

K⁺

K⁺

Na⁺

K⁺

At rest, sodium-potassium pumps in a neuron's membrane keep a higher concentration of sodium ions outside the cell and a higher concentration of potassium ions inside, creating a voltage difference called the resting potential. When an impulse moves down an axon, sodium ions rush into the cell, creating a reversal in voltage called an action potential.

HOLT BIOSOURCES / Teaching Transparencies **195**

Physiology of a Nerve Impulse 195

Teaching Strategies

- Use this transparency when discussing the propagation of a nerve impulse. Review the terms *resting potential*, *action potential*, and *repolarization*.
- Guide students through the events involved in nerve impulse propagation, beginning with the patch of membrane at resting potential.
- Emphasize that the concentration of sodium is higher outside the cell than inside the cell. Point out that as the nerve impulse travels along the axon of one neuron and down the length of the other neuron, sodium ions rush into the cell, temporarily reversing the voltage in a small patch of membrane.
- Inform students that local anesthetics act to eliminate pain by blocking the sodium gates in a neuron's cell membrane. Because the sodium gates do not open, an action potential cannot be generated, and a pain message cannot be sent to the brain.

BIOSOURCES
TRANSPARENCY MASTER

Major Endocrine Glands and Hormones, continued **195A**

Gland	Hormone	Target Tissue	Effects
Parathyroids	Parathyroid hormone	Bone tissue, digestive tract, kidneys	Stimulates breakdown of bone tissue and absorption of calcium by kidneys; raises blood calcium; activates vitamin D
Pineal	Melatonin	Uncertain, possibly ovaries and testes	May regulate biorhythms and moods; may affect the onset of puberty
Thyroid	Calcitonin	Bone tissue	Inhibits loss of calcium from bone; lowers blood calcium
	Thyroxine	All tissues	Raises metabolic rate; necessary for normal growth and development
Ovaries	Estrogen	All tissues, female reproductive structures	Controls development of secondary female sex characteristics and sex organs; initiates preparation of the uterus for pregnancy
	Progesterone	Uterus, breasts	Completes preparation of the uterus for pregnancy; stimulates breast development
Testes	Testosterone	All tissues, male reproductive structures	Controls development of secondary male sex characteristics and sex organs; stimulates sperm formation

195A HOLT BIOSOURCES / Teaching Transparencies

Major Endocrine Glands and Hormones, continued 195A

Teaching Strategies

- Use this table to review the human body's major glands and the hormones they produce.
- Ask students to identify the target organs for each hormone and the effects each hormone has on its target organ(s).

Which hormones control the development of secondary sex characteristics?
(estrogen and testosterone)

Which hormones regulate the body's blood-calcium level?
(parathyroid hormone and calcitonin)

Which hormones help regulate blood sugar?
(cortisol, epinephrine, glucagon, and insulin)

Peripheral Nervous System

Teaching Strategies

- Use this transparency when discussing the components of the peripheral nervous system. Review the terms *sensory neuron* and *motor neuron*.
- Identify the brain and the spine in the figure, and point out that peripheral nerves connect the brain and the spinal cord to the periphery of the body. Tell students that 12 pairs of nerves arise from the brain and 31 pairs of nerves arise from the spine.
- Emphasize that the peripheral nervous system is organized into two parts—the somatic nervous system, which innervates skeletal muscles, and the autonomic nervous system, which innervates smooth and cardiac muscles.

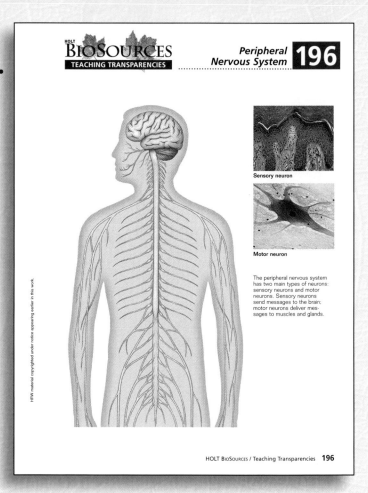

Sensory neuron

Motor neuron

The peripheral nervous system has two main types of neurons: sensory neurons and motor neurons. Sensory neurons send messages to the brain; motor neurons deliver messages to muscles and glands.

HRW material copyrighted under notice appearing earlier in this work.

HOLT BioSources / Teaching Transparencies **196**

Events of Human Fetal Development

Teaching Strategies

- Use this table to review the events that occur during human fetal development. Have students use the table to answer questions about human development.
- Ask students during which stage the basic body plan of a human is established *(the first trimester)*.

During which stages can fetal movements be felt by the mother?

(the second and third trimesters)

During which stage would a mutagenic agent be most likely to disrupt the normal development of a fetus?

(the first trimester)

196A
HOLT BioSources / Teaching Transparencies

HOLT BIOSOURCES
TRANSPARENCY MASTER

Events of Human Fetal Development 196A

Stage	Major Events
First Trimester (0 to 3 months)	Fertilization, cleavage, implantation, gastrulation, neurulation, and organogenesis occur as the embryo becomes a fetus; all major organ systems are formed; the fetus begins to move, but movements cannot be felt
Second Trimester (4 to 6 months)	Skin and hair grow; eyes blink; fetal movements can be felt; arms and legs reach final proportions; heartbeat can be heard
Third Trimester (7 to 9 months)	Substantial increase in size; skin is red and wrinkled; development of the lungs is completed; fingernails and toenails grow; fetus can survive if born during this stage

HRW material copyrighted under notice appearing earlier in this work.

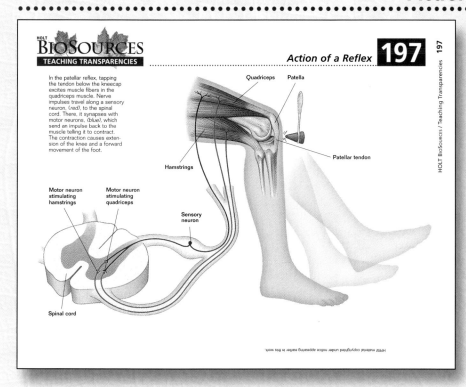

BIOSOURCES
TEACHING TRANSPARENCIES

Action of a Reflex **197**

In the patellar reflex, tapping the tendon below the kneecap excites muscle fibers in the quadriceps muscle. Nerve impulses travel along a sensory neuron, (*red*), to the spinal cord. There, it synapses with motor neurons, (*blue*), which send an impulse back to the muscle telling it to contract. The contraction causes extension of the knee and a forward movement of the foot.

Quadriceps   Patella

Hamstrings

Patellar tendon

Motor neuron stimulating hamstrings

Motor neuron stimulating quadriceps

Sensory neuron

Spinal cord

Teaching Strategies

- Use this transparency to illustrate the simple neural circuit of a reflex arc. Review the terms *sensory neuron* and *motor neuron*.

- Guide students through the events involved in the patellar reflex, beginning with the rubber-tipped hammer tapping the region below the knee cap. Explain that this tap stimulates muscle fibers in the quadriceps muscle. Sensory neurons carry information to the spinal cord, and motor neurons carry information back to the muscle.

- Inform students that doctors regularly test for the knee-jerk reflex to ascertain whether certain nerves and a portion of the spinal cord are functioning normally.

Divisions of the Nervous System 197A

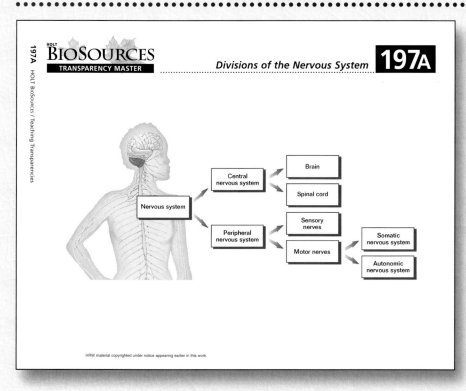

197A

HOLT
BIOSOURCES
TRANSPARENCY MASTER

Divisions of the Nervous System **197A**

Nervous system

Central nervous system

Peripheral nervous system

Brain

Spinal cord

Sensory nerves

Motor nerves

Somatic nervous system

Autonomic nervous system

Teaching Strategies

- Use this transparency to summarize the principal parts of the nervous system.

- Emphasize the two principal parts—the central nervous system and the peripheral nervous system. Explain that the peripheral nervous system includes two other divisions—the somatic nervous system and the autonomic nervous system.

- Ask students which nervous-system division gathers information about the environment and delivers it to the brain (*sensory nerves in the peripheral nervous system*).

Teaching Strategies

- Use this transparency to illustrate five types of joints that occur in the human skeleton.

- Review each type of joint illustrated in the figure. Demonstrate the movement made possible by the joints. Ask students to explain the advantages of having a skull that is composed of a patchwork of flat bones fused together with suture joints. *(This structure enables the skull of an infant to pass through the birth canal and allows the dome-shaped skull to grow as the body develops.)*

- Emphasize that in jointed skeletons, such as the human skeleton, skeletal muscles are attached to the different bones that make the joint. When the muscle contracts, it causes the joint between its two points of attachment to bend or straighten, depending on the placement of the muscle at the joint.

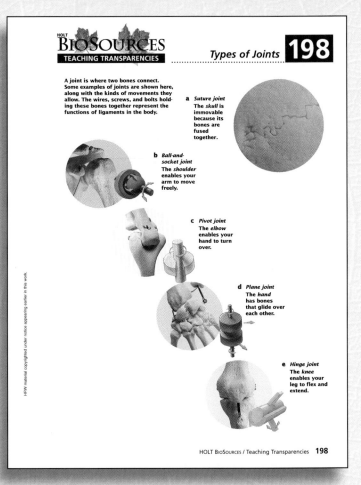

A joint is where two bones connect. Some examples of joints are shown here, along with the kinds of movements they allow. The wires, screws, and bolts holding these bones together represent the functions of ligaments in the body.

a *Suture joint* The *skull* is immovable because its bones are fused together.

b *Ball-and-socket joint* The *shoulder* enables your arm to move freely.

c *Pivot joint* The *elbow* enables your hand to turn over.

d *Plane joint* The *hand* has bones that glide over each other.

e *Hinge joint* The *knee* enables your leg to flex and extend.

HOLT BioSources / Teaching Transparencies **198**

198A *Structure of Representative Hormones*

Teaching Strategies

- Use this transparency to help students distinguish between peptide and steroid hormones.

- Point out that most peptide hormones, such as ADH, are chains of amino acids, comparable in size to an epinephrine molecule. Thus, most peptide hormones are quite large.

- Point out that each point on a five- or six-sided ring represents a carbon atom, while each black line represents a bond.

- Have students state the number of carbon atoms in a testosterone molecule *(17)* and in an epinephrine molecule *(9)*.

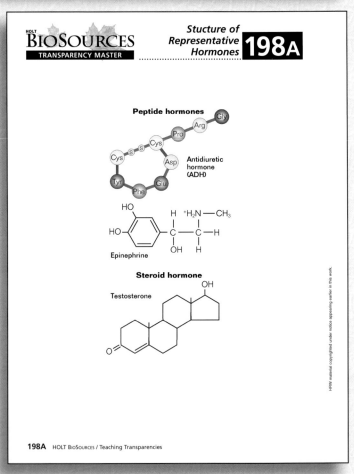

Peptide hormones

Antidiuretic hormone (ADH)

Epinephrine

Steroid hormone

Testosterone

198A HOLT BioSources / Teaching Transparencies

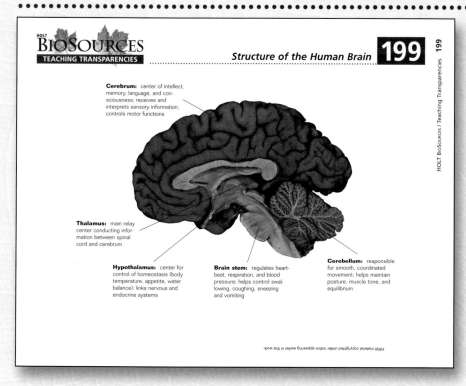

Teaching Strategies

- Use this transparency when discussing the major structures of the human brain. Read the names of the labeled structures aloud so that students hear correct pronunciation.
- Emphasize that the cerebrum consists of right and left hemispheres, which are connected internally by the corpus callosum. The inner portion of the cerebrum consists of white matter, and the outer portion is composed of gray matter, which is 2–4 mm thick in humans.
- Inform students that the limbic system, the part of the brain involved in behavioral responses and in memory, is not a distinct structure, but rather a group of regions that are related functionally.

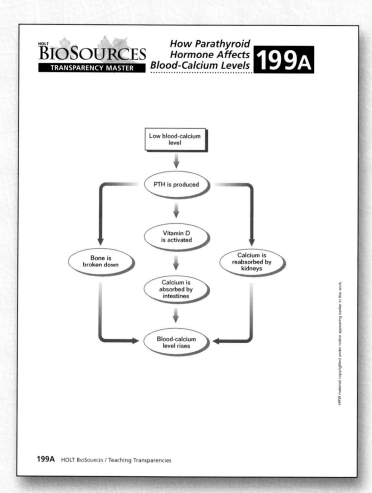

How Parathyroid Hormone Affects Blood-Calcium Levels 199A

Teaching Strategies

- Use this transparency when reviewing the three pathways by which PTH raises the blood-calcium level.
- Point out that PTH stimulates the breakdown of bone, which releases calcium into the bloodstream. Also point out that PTH stimulates the kidneys to reabsorb calcium into the bloodstream. Finally, add that by activating vitamin D, PTH indirectly stimulates the intestines to absorb calcium from digested food.
- Have students identify how the production of PTH might be regulated (by negative feedback).

Teaching Strategies

- Use this transparency when discussing the structures and functions of the inner ear. Review the terms *equilibrium, semicircular canals, otolith,* and *cochlea.*

- Review how structures in the inner ear sense equilibrium. Emphasize the role gravity plays in an individual's sense of equilibrium. *(Because gravity pulls objects downward, a change in the head's position or speed of motion causes otoliths to move and exert more pressure on some hair cells than on others.)*

- Explain how structures in the inner ear sense sound. Point out that the cuplike shape of the outer ear, the pinnae, captures sound vibrations in the air and funnels them into the auditory canal.

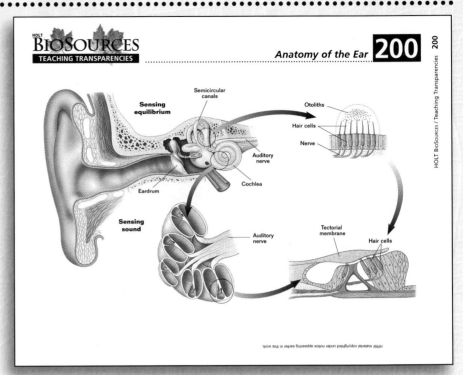

Teaching Strategies

- Use this transparency to illustrate how aldosterone helps maintain homeostasis by regulating body fluids and electrolytes.

- Tell students that aldosterone is very important to their health because it helps maintain the balance of sodium and potassium, both of which are crucial to nerve transmission. Emphasize aldosterone's importance to survival.

- Ask students to determine what triggers the release of aldosterone *(a loss of body fluids and the resulting drop in blood volume and blood pressure).*

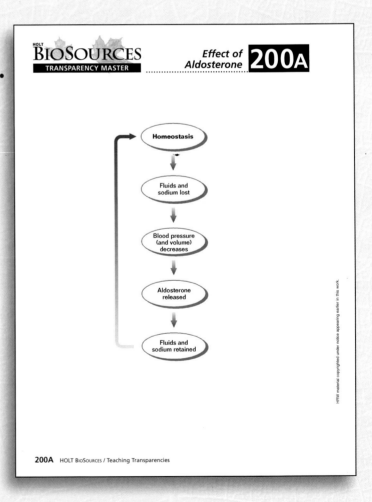

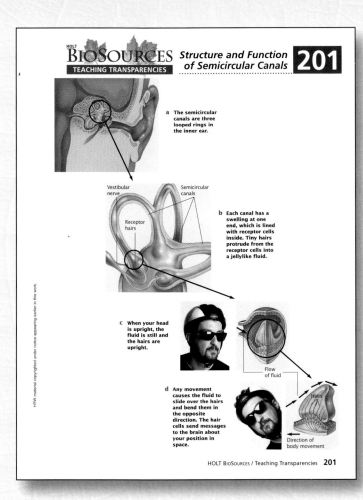

a The semicircular canals are three looped rings in the inner ear.

Vestibular nerve Semicircular canals

Receptor hairs

b Each canal has a swelling at one end, which is lined with receptor cells inside. Tiny hairs protrude from the receptor cells into a jellylike fluid.

c When your head is upright, the fluid is still and the hairs are upright.

Flow of fluid

d Any movement causes the fluid to slide over the hairs and bend them in the opposite direction. The hair cells send messages to the brain about your position in space.

Hairs

Direction of body movement

HOLT BioSources / Teaching Transparencies **201**

Teaching Strategies

- Use this transparency to illustrate the structure of the semicircular canals.
- Point out that each of the three canals is oriented at a right angle to the other two. This design makes it possible to sense rotational acceleration (changes in the rate or direction of rotation of the head).
- Inform students that when the head is rotated for an extended period and then stopped abruptly, the fluid in the canals will continue to circulate. The result is the kind of dizziness experienced after some amusement park rides.

Types of Psychoactive Drugs | 201A

Drug	Examples	Effects	Risks Associated With Abuse
Depressants	Barbiturates, tranquilizers, methaqualone, phencyclidine hydrochloride (PCP)	Slow down the action of the central nervous system	Drowsiness, depression, emotional instability
Stimulants	Amphetamines, caffeine	Speed up the central nervous system, metabolism, blood pressure, heartbeat, and respiratory rate	Irregular heartbeat, high blood pressure, headaches, stomach disorders, exhaustion, violent behavior
Inhalants	Nitrous oxide, ether, paint thinners, glue, cleaning fluids, correction fluids	Disorientation, confusion, memory loss	Hallucinations; permanent damage to brain, kidneys, liver; death
Hallucinogens	Lysergic acid diethylamide (LSD), mescaline, peyote	Distortion in the way the brain translates impulses	Dangerous hallucinations, unpredictable behavior
Marijuana	Derived from dried leaves, flowers, and stems of Cannabis sativa	Wide range of effects; short-term memory loss, disorientation, impaired judgment	Lung damage, loss of motivation

HRW material copyrighted under notice appearing earlier in this work.

Teaching Strategies

- Use this transparency to summarize five major types of psychoactive drugs. Review the term *psychoactive drug*.
- Review each type of drug listed in the table, citing examples, effects, and risks associated with the use and abuse of each.

Teaching Strategies

- Use this transparency when discussing the structure of the human eye. Review the major structures involved in vision and the terms *rod* and *cone*.

- Trace the path light travels through the eye. Inform students that the human eye is a camera-type eye. It uses a single lens to focus light on a surface containing many photoreceptors. In this type of eye, the retina functions in a manner analogous to a piece of film.

- Review the defects that occur in the lens to cause impairments such as near-sighted vision and farsighted vision. Explain that a common defect known as astigmatism stems from uneven curvature of the cornea.

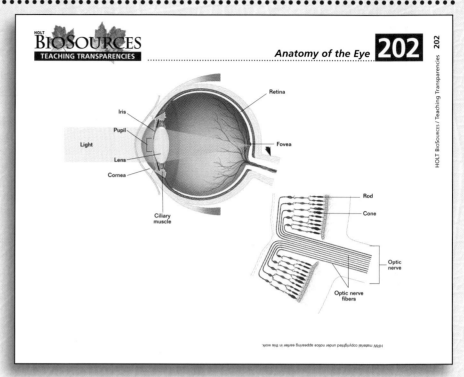

202A *Immune System Response Time*

Teaching Strategies

- Use this transparency when discussing immune-system response time in the human body. Review the terms *antibody* and *pathogen*.

- Point out the importance of memory cells. These cells are produced when the body first encounters a pathogen. At any subsequent encounter with the same pathogen, memory cells quickly divide to produce a large number of antibody-secreting cells, and the pathogen is defeated before it can cause disease.

- Have students explain how vaccination makes use of memory cells. *(A vaccine contains pathogens or toxoids whose disease-causing ability has been removed. A vaccine stimulates the production of antibodies and memory cells within the immune system.)*

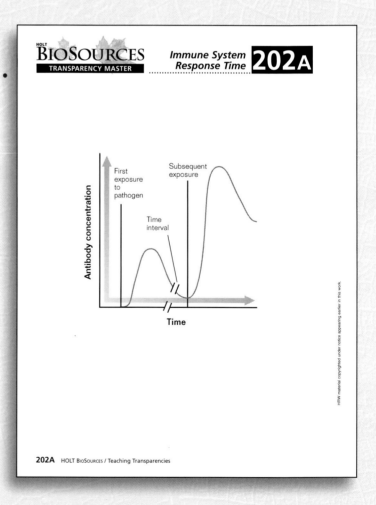

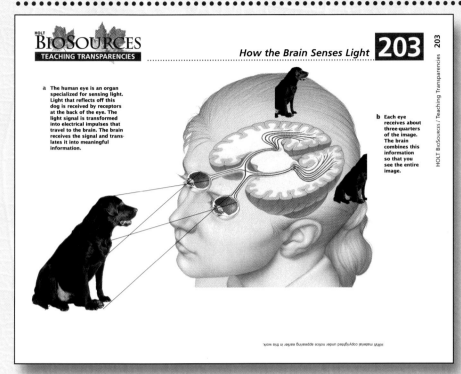

Teaching Strategies

- Use this transparency when explaining how the brain senses light. Review the terms *cornea, retina*, and *refraction*.

- Trace the path light travels through the human eye, beginning with the light that reflects from an object in the environment (the dog). Emphasize that the light is bent by the cornea and is focused on the retina.

- Inform students that the image that forms on the retina is inverted.

Where is a visual image converted to reflect an object's true orientation?

(in the brain)

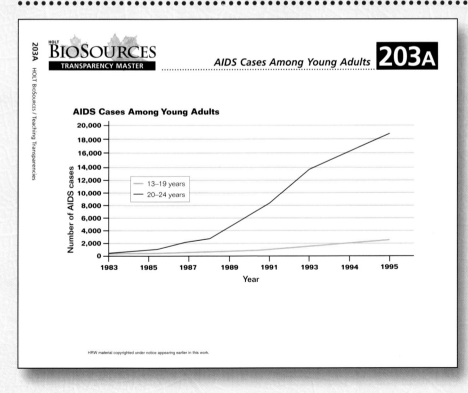

Teaching Strategies

- Use this transparency to illustrate how the number of reported AIDS cases in 13- to 19-year-olds and 20- to 24-year-olds rose from 1983 to 1995. Emphasize that these are cumulative figures, compiled since 1983. Have students compare the rise in the graph lines for the two groups.

- Have students determine when a 20-year-old person with AIDS most likely contracted HIV *(during his or her teens)*.

204 *Endocrine System*

Teaching Strategies

- Use this transparency when discussing the major human endocrine glands and organs that produce hormones. Review the term *hormone*. Explain that hormones are specific chemical messengers that produce an effect away from their sites of production.

- Explain that hormones have been found in a wide variety of invertebrates, such as arthropods, annelids, mollusks, and echinoderms, but that far more is known about hormones in mammals, particularly humans.

Why are endocrine glands commonly called the ductless glands?

(They secrete hormones directly into the blood via the capillaries. No special ducts are involved.)

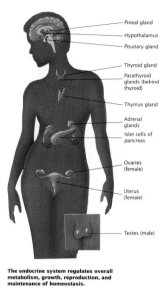

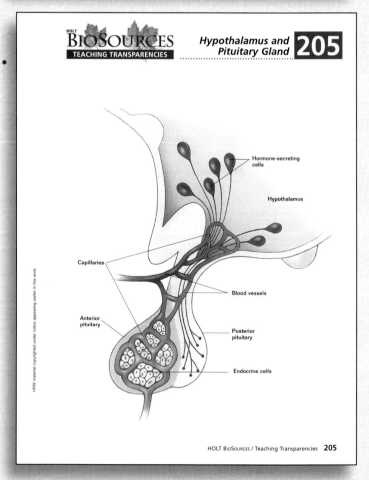

---

205 *Hypothalamus and Pituitary Gland*

Teaching Strategies

- Use this transparency when discussing the structure of the hypothalamus and the pituitary gland.

- Have students compare the internal structure of the two lobes of the pituitary gland. Tell students that blood vessels (not shown) enter the posterior lobe and distribute the hormones stored there throughout the body.

- Ask students to explain what the anterior lobe contains *(many endocrine cells and blood vessels)*.

How does this structure relate to the function of the anterior lobe?

(Hormones are produced in this lobe and enter the bloodstream, which distributes them throughout the body.)

Action of a Steroid Hormone

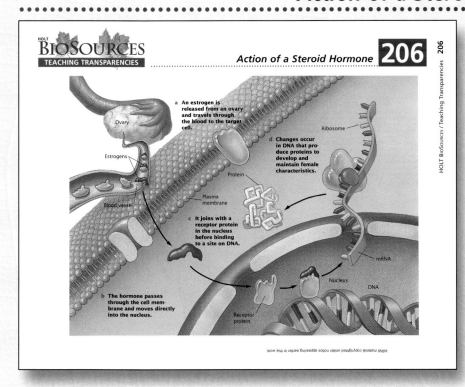

Teaching Strategies

- Use this transparency to illustrate the action of the steroid hormone estrogen. Review the characteristics of the cell membrane.

- Walk students through the steps in this diagram. Point out that estrogen is a steroid hormone that enters a cell and binds with a receptor protein inside the cell. Tell students that estrogen stimulates the development and maintenance of female secondary sexual characteristics and reproductive structures. It also stimulates the growth of the uterine lining.

- Have students explain what happens after estrogen binds to its receptor protein. *(The hormone-receptor complex enters the cell's nucleus and activates genes that produce certain proteins.)*

Action of a Peptide Hormone

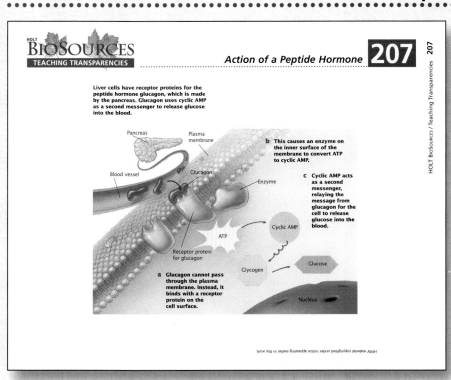

Teaching Strategies

- Use this transparency to illustrate the action of the peptide hormone glucagon. Review the terms *ATP* and *second messenger*.

- Walk students through the steps in this diagram. Tell students that glucagon is a peptide hormone that helps maintain homeostasis by stimulating the breakdown of glycogen into glucose, which is used for making ATP. Point out that the receptor protein for glucagon is in the cell membrane and that glucagon does not enter the cell.

- Have students explain what happens after glucagon binds to a receptor. *(Cyclic AMP, a second messenger, activates enzymes that convert glycogen to glucose.)*

 BioSources / Transparency Directory with Teacher's Notes **139**

Teaching Strategies

- Use this transparency when explaining how the body uses hormones to maintain normal blood glucose levels.

- Be sure students are aware that the increase in insulin secretion results in a lowering of blood glucose. When blood glucose drops too low, glucagon secretion is stimulated and insulin secretion is inhibited, and the blood glucose level rises.

- Inform students that the interaction between the secretion of these two hormones and the concentration of glucose in the blood is an example of negative-feedback control, a homeostatic strategy very common in living organisms.

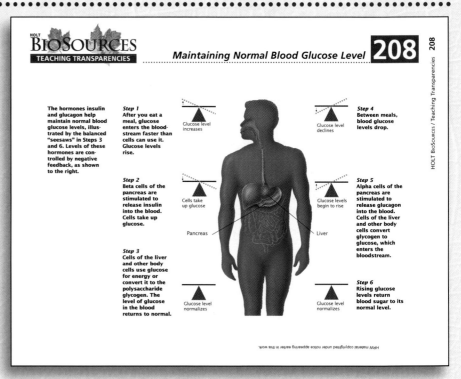

208 Maintaining Normal Blood Glucose Level

The hormones insulin and glucagon help maintain normal blood glucose levels, illustrated by the balanced "seesaws" in Steps 3 and 6. Levels of these hormones are controlled by negative feedback, as shown to the right.

Step 1 After you eat a meal, glucose enters the bloodstream faster than cells can use it. Glucose levels rise.

Step 2 Beta cells of the pancreas are stimulated to release insulin into the blood. Cells take up glucose.

Step 3 Cells of the liver and other body cells use glucose for energy or convert it to the polysaccharide glycogen. The level of glucose in the blood returns to normal.

Step 4 Between meals, blood glucose levels drop.

Step 5 Alpha cells of the pancreas are stimulated to release glucagon into the blood. Cells of the liver and other body cells convert glycogen to glucose, which enters the bloodstream.

Step 6 Rising glucose levels return blood sugar to its normal level.

Glucose level increases — Cells take up glucose — Glucose level normalizes — Pancreas — Glucose level declines — Glucose levels begin to rise — Glucose level normalizes — Liver

Teaching Strategies

- Use this transparency to show students where the major endocrine glands are located in the human body.

- Point out that although the hypothalamus is part of the brain, it is also considered an endocrine gland because it produces many hormones. Tell students that several body organs with other functions, such as the heart, stomach, and kidneys, contain hormone-secreting cells and thus may also be considered endocrine glands.

- Ask students where the parathyroid glands are located. (*They are located behind the thyroid gland.*)

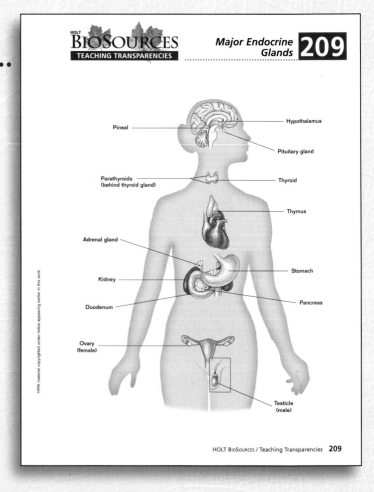

209 Major Endocrine Glands

Pineal — Hypothalamus — Pituitary gland — Parathyroids (behind thyroid gland) — Thyroid — Thymus — Adrenal gland — Stomach — Kidney — Pancreas — Duodenum — Ovary (female) — Testicle (male)

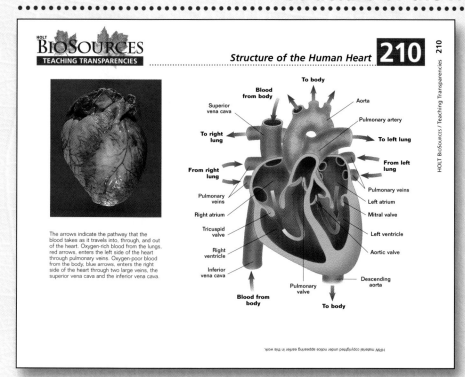

Structure of the Human Heart 210

HOLT
BIOSOURCES
TEACHING TRANSPARENCIES

To body

Blood from body

Superior vena cava

Aorta

Pulmonary artery

To right lung

To left lung

From right lung

From left lung

Pulmonary veins

Pulmonary veins

Left atrium

Right atrium

Mitral valve

Tricuspid valve

Left ventricle

Right ventricle

Aortic valve

Inferior vena cava

Descending aorta

Pulmonary valve

To body

Blood from body

The arrows indicate the pathway that the blood takes as it travels into, through, and out of the heart. Oxygen-rich blood from the lungs, red arrows, enters the left side of the heart through pulmonary veins. Oxygen-poor blood from the body, blue arrows, enters the right side of the heart through two large veins, the superior vena cava and the inferior vena cava.

Teaching Strategies

- Use this transparency to illustrate the structure of the human heart. Review the terms *artery, vein, atrium,* and *ventricle*.
- Trace the pathway of blood through the heart with your students. Point out that the wall of the left ventricle is thicker than the wall of the right ventricle. This is because the left ventricle must pump blood throughout the entire body. The right ventricle pumps blood only to the lungs.

Why is the wall of the aorta thicker than the wall of any other artery in the body?

(The blood inside the aorta comes directly from the left ventricle and is under very high pressure.)

A Closer Look at Blood Vessels 211

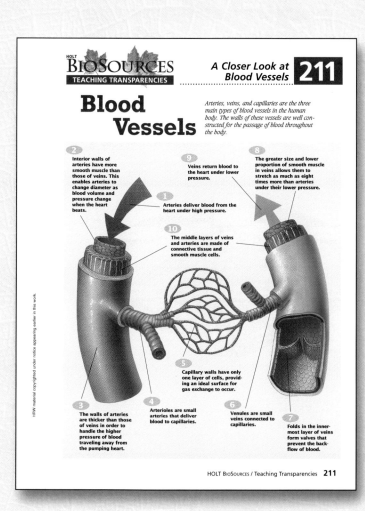

HOLT
BIOSOURCES
TEACHING TRANSPARENCIES

A Closer Look at Blood Vessels **211**

Blood Vessels

Arteries, veins, and capillaries are the three main types of blood vessels in the human body. The walls of these vessels are well constructed for the passage of blood throughout the body.

2 Interior walls of arteries have more smooth muscle than those of veins. This enables arteries to change diameter as blood volume and pressure change when the heart beats.

9 Veins return blood to the heart under lower pressure.

8 The greater size and lower proportion of smooth muscle in veins allows them to stretch as much as eight times more than arteries under their lower pressure.

1 Arteries deliver blood from the heart under high pressure.

10 The middle layers of veins and arteries are made of connective tissue and smooth muscle cells.

3 The walls of arteries are thicker than those of veins in order to handle the higher pressure of blood traveling away from the pumping heart.

4 Arterioles are small arteries that deliver blood to capillaries.

5 Capillary walls have only one layer of cells, providing an ideal surface for gas exchange to occur.

6 Venules are small veins connected to capillaries.

7 Folds in the innermost layer of veins form valves that prevent the backflow of blood.

HOLT BIOSOURCES / Teaching Transparencies **211**

A Closer Look at Blood Vessels 211

Teaching Strategies

- Use this transparency to illustrate the structure of blood vessels in the human body. Review the three main types of blood vessels.
- Point out that arteries and veins are composed of three layers. Have students describe each layer and its function. *(An outer layer of connective tissue lends flexibility; a middle layer of smooth muscle changes the size of the vessels; and a layer of endothelium adds a smooth, protective inner membrane.)*
- Be sure students are aware that arteries and veins have the same three layers but that the walls of veins are thinner and less rigid and readily change shape when muscles press against them.

Teaching Strategies

- Use this transparency to illustrate circulation through the human body. Be sure students are aware that red vessels contain oxygenated blood and that blue vessels contain deoxygenated blood.
- Trace the path of systemic circulation through the body, beginning with blood flowing through the right ventricle to the lungs. Ask students what happens to the blood when it passes through the lungs. (*It picks up oxygen.*)
- Have students design a graphic organizer that compares systemic circulation in humans with that in fish, amphibians, and reptiles.

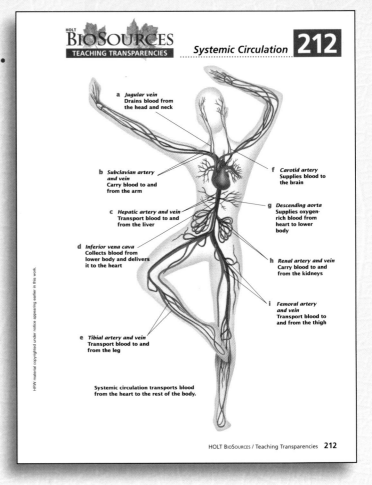

HOLT BioSources
TEACHING TRANSPARENCIES

Systemic Circulation **212**

a **Jugular vein** Drains blood from the head and neck

b **Subclavian artery and vein** Carry blood to and from the arm

c **Hepatic artery and vein** Transport blood to and from the liver

d **Inferior vena cava** Collects blood from lower body and delivers it to the heart

e **Tibial artery and vein** Transport blood to and from the leg

f **Carotid artery** Supplies blood to the brain

g **Descending aorta** Supplies oxygen-rich blood from heart to lower body

h **Renal artery and vein** Carry blood to and from the kidneys

i **Femoral artery and vein** Transport blood to and from the thigh

Systemic circulation transports blood from the heart to the rest of the body.

HRW material copyrighted under notice appearing earlier in this work.

HOLT BioSources / Teaching Transparencies **212**

213 *Circulation Pathway in the Human Body*

Teaching Strategies

- Use this transparency to illustrate the pathway blood travels through the human body. Review the various types of blood vessels.
- Remind students that humans have a closed circulatory system, meaning the blood never leaves the blood vessels and never bathes the body's tissues directly. Substances in the blood that are needed by tissues, like oxygen and food molecules, must pass through the walls of capillaries.
- Have students describe the basic structure and function of the human heart. (*It is a four-chambered muscular pump that propels blood through the blood vessels of the circulatory system.*)

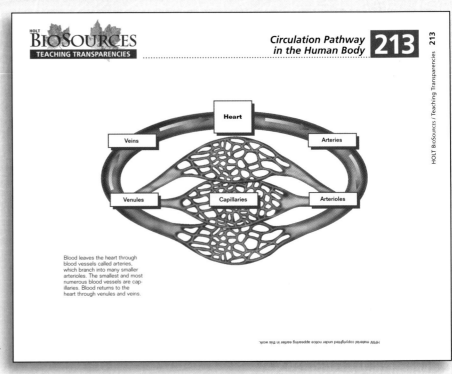

HOLT BioSources
TEACHING TRANSPARENCIES

Circulation Pathway in the Human Body **213**

Heart

Veins

Arteries

Venules

Capillaries

Arterioles

Blood leaves the heart through blood vessels called arteries, which branch into many smaller arterioles. The smallest and most numerous blood vessels are capillaries. Blood returns to the heart through venules and veins.

HRW material copyrighted under notice appearing earlier in this work.

HOLT BioSources / Teaching Transparencies **213**

HOLT
BIOSOURCES
TEACHING TRANSPARENCIES

Circulatory Loops in the Human Body **214**

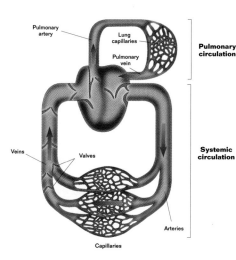

Pulmonary artery

Lung capillaries

Pulmonary vein

Pulmonary circulation

Veins

Valves

Systemic circulation

Arteries

Capillaries

The pulmonary circulatory loop transports blood from the right side of the heart to the lungs and then to the left side of the heart. The systemic circulatory loop transports blood from the left side of the heart to all the body's tissues and then to the right side of the heart. The designation of the left and right sides of the heart in the diagram is based on the left and right sides of an intact heart that is still in a body.

HRW material copyrighted under notice appearing earlier in this work.

Circulatory Loops in the Human Body 214

Teaching Strategies

- Use this transparency to illustrate the circulatory loops blood travels through the human body.
- Remind students that in almost all cases, the blood inside arteries is oxygen-rich and the blood inside veins is oxygen-poor. The only exception to this rule is the blood inside the pulmonary artery and pulmonary vein. The pulmonary artery contains oxygen-poor blood, and the pulmonary vein contains oxygen-rich blood.
- Ask students to explain what the capillaries at the bottom of the figure represent. *(They represent all capillaries in all tissues of the body.)*
- Ask students to develop a rule to help them remember a major difference between arteries and veins. *(Arteries carry blood away from the heart. Veins carry blood toward the heart.)*

HOLT
BIOSOURCES
TEACHING TRANSPARENCIES

Respiratory System in the Human Body **215**

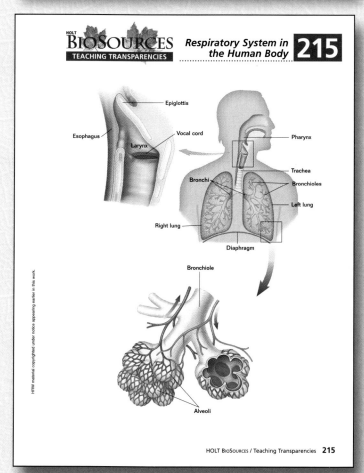

Epiglottis

Esophagus

Vocal cord

Larynx

Pharynx

Trachea

Bronchi

Bronchioles

Left lung

Right lung

Diaphragm

Bronchiole

Alveoli

HRW material copyrighted under notice appearing earlier in this work.

Respiratory System in the Human Body 215

Teaching Strategies

- Use this transparency to illustrate the pathway air travels through the human body.
- Trace the path of air through the passages of the respiratory system shown in the diagram. Remind students that the exchange of oxygen and carbon dioxide in the lungs occurs across the walls of alveoli, such as those shown in the figure.

How are oxygen and carbon dioxide exchanged in the lungs?

(Oxygen diffuses from the alveoli into the surrounding capillaries, and carbon dioxide diffuses from the capillaries into the alveoli.)

Teaching Strategies

- Use this transparency when discussing the major organs of the human digestive system.
- Remind students that digestion begins in the mouth, where food is chewed and chemically digested by amylase, which is produced in the salivary glands.
- Point out the liver, pancreas, and gallbladder. Food never enters these organs. Instead, they secrete various substances into the digestive tract.
- Have students describe what peristaltic contractions are. *(They are the rhythmic waves of contraction of the smooth muscle in the wall of the esophagus that move food along toward the stomach.)*

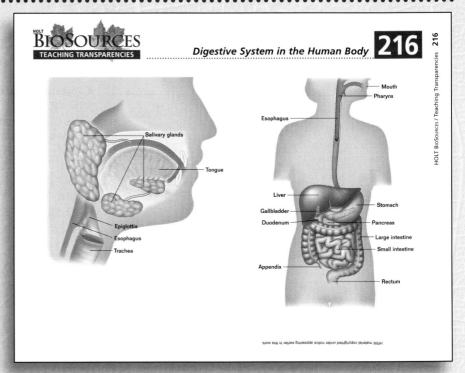

217 *Overview of Respiration*

Teaching Strategies

- Use this transparency when discussing respiration. Point out the directions that oxygen and carbon dioxide move in the lungs and in the body's tissues.
- Remind students that 70 percent of the carbon dioxide that travels to the lungs is in the form of bicarbonate ions. Review how a bicarbonate ion is made from carbon dioxide inside red blood cells.
- Have students explain how oxygen is transported in the blood. *(Oxygen binds to the hemoglobin molecules inside red blood cells.)*

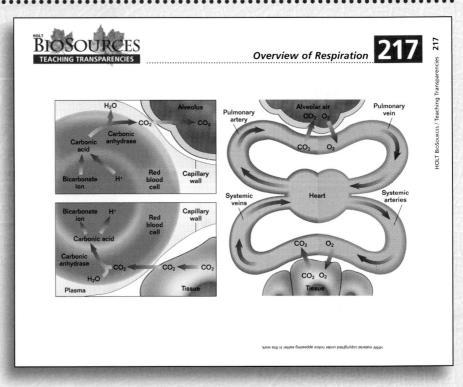

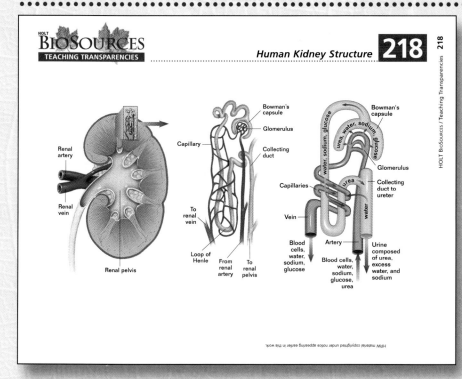

HOLT BIOSOURCES
TEACHING TRANSPARENCIES

Human Kidney Structure 218

HOLT BioSources / Teaching Transparencies 218

Renal artery

Renal vein

Renal pelvis

Capillary

Bowman's capsule

Glomerulus

Collecting duct

To renal vein

From renal artery

To renal pelvis

Loop of Henle

Blood cells, water, sodium, glucose

water, sodium, glucose

urea water, sodium, glucose

Bowman's capsule

Glomerulus

urea

Capillaries

Vein

Artery

water

Collecting duct to ureter

Blood cells, water, sodium, glucose

Urine composed of urea, excess water, and sodium

HRW material copyrighted under notice appearing earlier in this work.

Teaching Strategies

- Use this transparency when describing the structure of a human kidney. Review the terms *artery*, *vein*, and *nephron*.
- Tell students that each kidney contains roughly 1 million nephrons.
- Point out the parts of a nephron. Review the figure on the far right to help explain how the processes of filtration and reabsorption lead to the formation of urine.

What is urea and where is it made?

(Urea is a nitrogenous waste produced in the liver.)

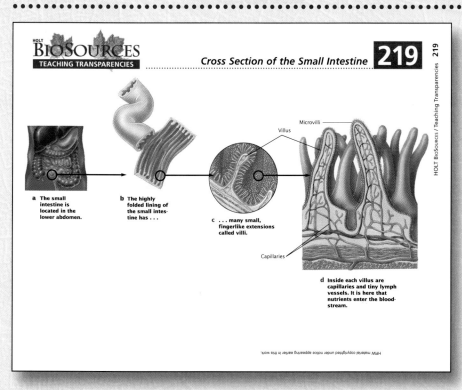

HOLT BIOSOURCES
TEACHING TRANSPARENCIES

Cross Section of the Small Intestine 219

HOLT BioSources / Teaching Transparencies 219

Microvilli

Villus

a The small intestine is located in the lower abdomen.

b The highly folded lining of the small intestine has . . .

c . . . many small, fingerlike extensions called villi.

Capillaries

d Inside each villus are capillaries and tiny lymph vessels. It is here that nutrients enter the bloodstream.

HRW material copyrighted under notice appearing earlier in this work.

Teaching Strategies

- Use this transparency when discussing digestion in the human body. Review the term *villi*.
- Point out that most digestion and absorption take place in the small intestine. Here carbohydrates, fats, and proteins are hydrolyzed by enzymes into the final products of digestion—simple sugars, fatty acids, and amino acids—that are absorbed and transported by the blood to all the cells of the body.
- Point out that the intestinal lining has folds bearing numerous villi, which greatly increase the absorptive surface area. The villi have smooth muscle fibers that enable them to move back and forth. Such movement increases after a meal.

220 How a Kidney Machine Works

Teaching Strategies

- Use this transparency to illustrate how dialysis machines work to artificially clean wastes from the blood.
- Review the term *semipermeable membrane*. Have students explain why this membrane is essential for the dialysis process.

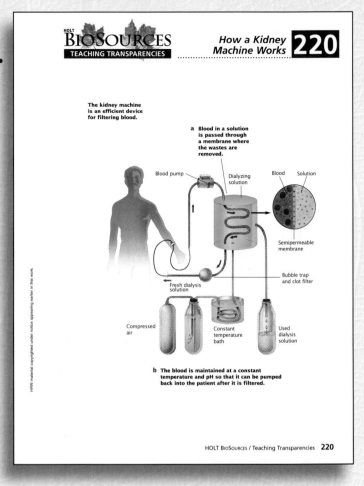

The kidney machine is an efficient device for filtering blood.

a Blood in a solution is passed through a membrane where the wastes are removed.

Blood pump

Dialyzing solution

Blood   Solution

Semipermeable membrane

Fresh dialysis solution

Bubble trap and clot filter

Compressed air

Constant temperature bath

Used dialysis solution

b The blood is maintained at a constant temperature and pH so that it can be pumped back into the patient after it is filtered.

HOLT BIOSOURCES / Teaching Transparencies **220**

---

221 Excretory System in the Human Body

Teaching Strategies

- Use this transparency to illustrate the major organs of the human excretory system.
- Explain that the right kidney is slightly lower than the left because of the large amount of space taken up by the liver.
- Trace the path of urine from the kidneys through the ureter to the urinary bladder and through the urethra.
- Ask students how the cells that make up the kidneys get oxygen. (*They get oxygen from the blood that enters the kidneys through the renal arteries.*)

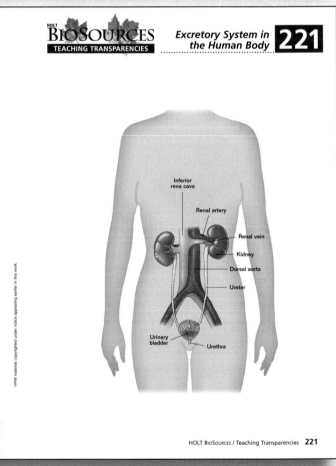

Inferior rena cava

Renal artery

Renal vein

Kidney

Dorsal aorta

Ureter

Urinary bladder

Urethra

HOLT BIOSOURCES / Teaching Transparencies **221**

BIOSOURCES
TEACHING TRANSPARENCIES
USDA Food Pyramid 222

The USDA food pyramid recommends the foods at the top of the pyramid be eaten sparingly, while the other foods may be eaten more often.

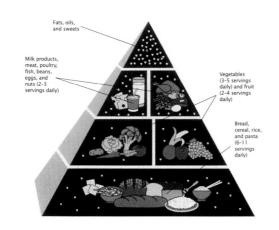

Fats, oils, and sweets

Milk products, meat, poultry, fish, beans, eggs, and nuts (2–3 servings daily)

Vegetables (3–5 servings daily) and fruit (2–4 servings daily)

Bread, cereal, rice, and pasta (6–11 servings daily)

USDA Food Pyramid 222

Teaching Strategies

- Use this transparency when discussing the requirements for a healthful diet.
- Walk students through the food pyramid, beginning at the base. Review the recommended daily servings for each food category.
- Emphasize that carbohydrates form the foundation of a balanced diet. Ask students to propose why carbohydrates are so important. *(They are sources of readily available energy as sugars and starches; they also provide dietary fiber, vitamins, and minerals.)*
- Discuss the importance of at least five servings of fruits and vegetables each day. Tell students that the chemical pigments provided by highly colored fruits and vegetables have important roles within cells.

BIOSOURCES
TEACHING TRANSPARENCIES
Female Reproductive System 223

The female reproductive system releases eggs and provides nourishment to a developing fetus.

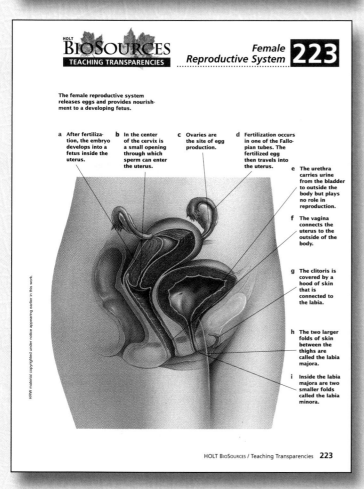

a After fertilization, the embryo develops into a fetus inside the uterus.

b In the center of the cervix is a small opening through which sperm can enter the uterus.

c Ovaries are the site of egg production.

d Fertilization occurs in one of the Fallopian tubes. The fertilized egg then travels into the uterus.

e The urethra carries urine from the bladder to outside the body but plays no role in reproduction.

f The vagina connects the uterus to the outside of the body.

g The clitoris is covered by a hood of skin that is connected to the labia.

h The two larger folds of skin between the thighs are called the labia majora.

i Inside the labia majora are two smaller folds called the labia minora.

Female Reproductive System 223

Teaching Strategies

- Use this transparency to review the structures and functions of the female reproductive system. Review the terms *uterus*, *ovary*, and *fallopian tube*.
- Review each structure of the female reproductive system. Tell students that during pregnancy, the uterus expands to a size that would fill the part of the abdomen seen in this diagram.
- Have students write a paragraph that describes the life story of an egg, detailing what happens as the egg develops and travels to the site of fertilization.
- Ask students to describe what happens to the mother's abdominal organs as a fetus grows. *(Her organs become compressed.)*

224 Male Reproductive System

Teaching Strategies

- Use this transparency to review the structures and functions of the male reproductive system. Review the term *testes*.
- Review the structure of the male reproductive system. Ask students to give examples of how the structure of the reproductive system helps the system accomplish its function.
- Have students write a paragraph describing the life of a sperm cell, from its production in the seminiferous tubules to its exit from the body.
- Ask students which tube carries sperm out of the body (*urethra*).

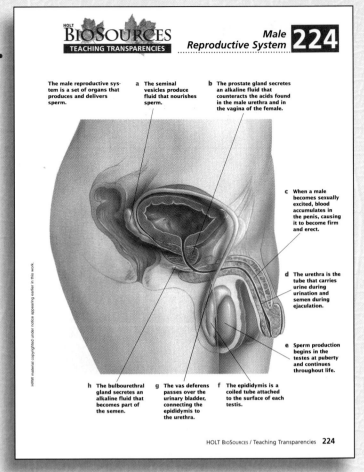

The male reproductive system is a set of organs that produces and delivers sperm.

a The seminal vesicles produce fluid that nourishes sperm.

b The prostate gland secretes an alkaline fluid that counteracts the acids found in the male urethra and in the vagina of the female.

c When a male becomes sexually excited, blood accumulates in the penis, causing it to become firm and erect.

d The urethra is the tube that carries urine during urination and semen during ejaculation.

e Sperm production begins in the testes at puberty and continues throughout life.

h The bulbourethral gland secretes an alkaline fluid that becomes part of the semen.

g The vas deferens passes over the urinary bladder, connecting the epididymis to the urethra.

f The epididymis is a coiled tube attached to the surface of each testis.

225 Male Hormones and Reproduction

Teaching Strategies

- Use this transparency when discussing the hormones involved in male reproduction. Review the term *testosterone*.
- Review the types of hormones involved in male reproduction. Emphasize that sperm production is controlled by levels of testosterone.
- Have students design a flowchart that illustrates how a negative-feedback system regulates levels of testosterone.

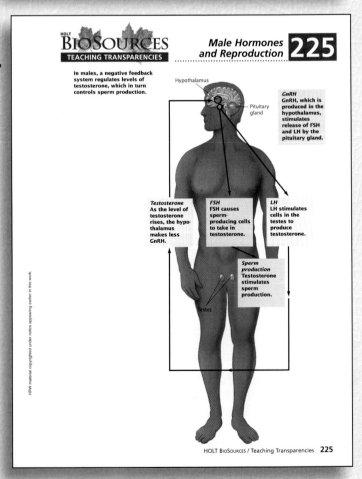

In males, a negative feedback system regulates levels of testosterone, which in turn controls sperm production.

Hypothalamus

Pituitary gland

GnRH
GnRH, which is produced in the hypothalamus, stimulates release of FSH and LH by the pituitary gland.

Testosterone
As the level of testosterone rises, the hypothalamus makes less GnRH.

FSH
FSH causes sperm-producing cells to take in testosterone.

LH
LH stimulates cells in the testes to produce testosterone.

Sperm production
Testosterone stimulates sperm production.

Testes

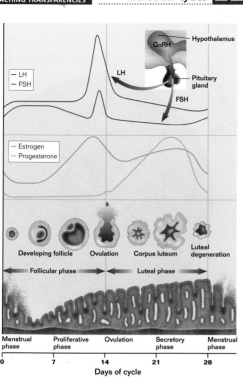

HOLT BioSources / Teaching Transparencies **226**

Teaching Strategies

- Use this transparency when comparing the ovarian and menstrual cycles. Review the terms *estrogen* and *ovulation*.
- Lead students through the events involved in each cycle. Point out the changes that occur in the ovary and in the uterine wall during these cycles.
- Ask students which hormones reach their peak production just before ovulation (*LH, FSH, and estrogen*).

What kind of mechanism keeps the uterus and the ovaries synchronized with each other?

(a negative-feedback mechanism)

Events Leading to Implantation 227

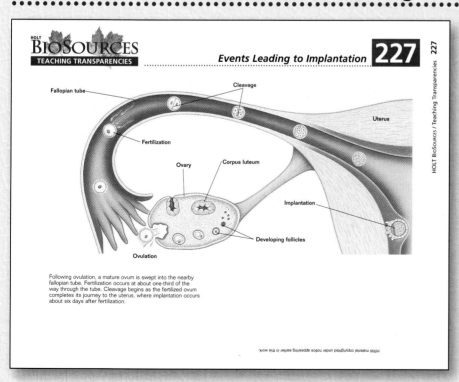

Following ovulation, a mature ovum is swept into the nearby fallopian tube. Fertilization occurs at about one-third of the way through the tube. Cleavage begins as the fertilized ovum completes its journey to the uterus, where implantation occurs about six days after fertilization.

Teaching Strategies

- Use this transparency when discussing the events that precede implantation. Review the terms *ovulation, fertilization, cleavage*, and *implantation*.
- Walk students through this figure, pointing out important stages in early embryonic development, such as fertilization, cleavage, and implantation. Also point out where these stages occur.
- Have students write a paragraph describing what happens to a mature ovum after it leaves the ovary and is fertilized by a sperm.
- Have students describe the process of cleavage. *(It is a series of cell divisions that increase the number of cells in the developing embryo without increasing its overall size.)*

Teaching Strategies

- Use this transparency when discussing how an embryo is attached to the uterine wall.

- Point out that the placenta forms from maternal and embryonic tissue. However, maternal and embryonic blood do not mix in the placenta. Wastes move from fetal to maternal blood, and nutrients move in the opposite direction, by diffusion. Inform students that drugs and pathogens can enter the embryo's blood through the placenta.

- Have students describe the function of the chorionic villi. *(They contain embryonic blood vessels that exchange wastes and nutrients with maternal blood.)*

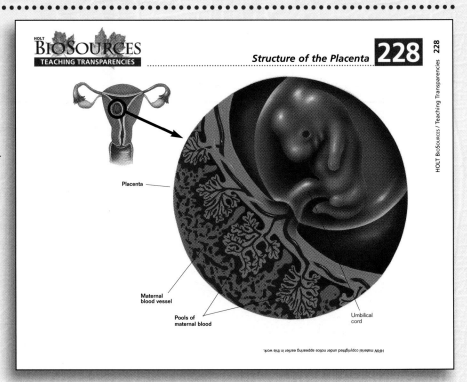

HOLT BIOSOURCES / TEACHING TRANSPARENCIES

Structure of the Placenta **228**

Placenta

Maternal blood vessel

Pools of maternal blood

Umbilical cord

Teaching Strategies

- Use this transparency to illustrate how hormones get to their target organs.

- Point out that the liver and pancreas are located very close to one another. Be sure students notice that the hormones produced in the pancreas enter the bloodstream, where they circulate through the body. In the liver, the hormones exit the bloodstream and cause a response in liver cells.

- Have students describe how the delivery of hormones differs from that of other chemicals. *(They are delivered to their target organs by the bloodstream.)*

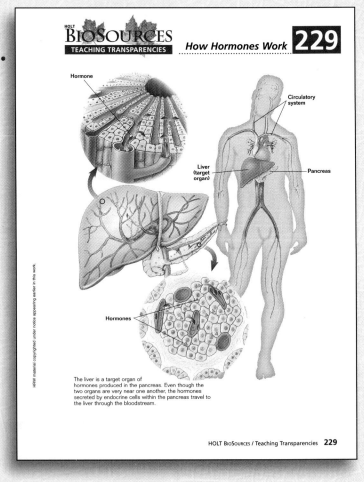

HOLT BIOSOURCES / TEACHING TRANSPARENCIES

How Hormones Work **229**

Hormone

Circulatory system

Liver (target organ)

Pancreas

Hormones

The liver is a target organ of hormones produced in the pancreas. Even though the two organs are very near one another, the hormones secreted by endocrine cells within the pancreas travel to the liver through the bloodstream.

HOLT BIOSOURCES / Teaching Transparencies **229**

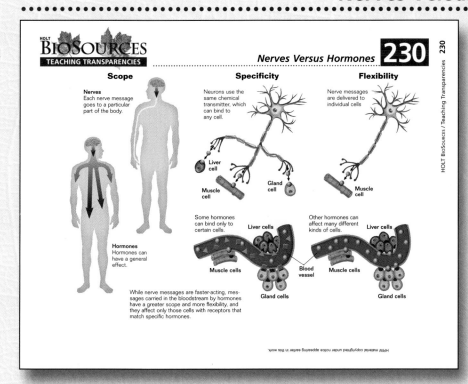

Teaching Strategies

- Use this transparency to compare the action of nerves with that of hormones.
- Walk students through the diagram. Point out that hormones are more flexible than neurons because one type of hormone can affect any cell that has a receptor for that hormone. Explain that hormones are more specific because they can attach only to cells with matching receptors, while all neurons use the same chemical transmitter.
- Ask students why hormones have greater scope than nerve messages. *(A neuron affects one part of the body, while one type of hormone can affect all parts.)*

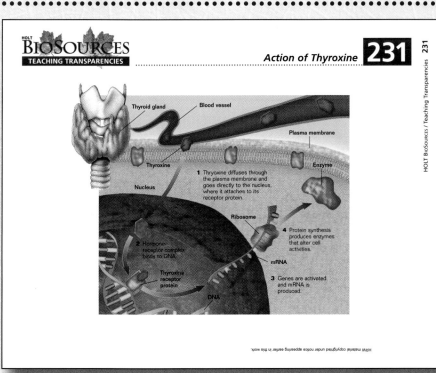

Teaching Strategies

- Use this transparency to illustrate the action of thyroxine.
- Walk students through the steps in this diagram. Point out that although thyroxine is a peptide hormone, it is made of only two amino acids and thus is small enough to pass through openings in a cell membrane. Tell students that thyroxine stimulates cell metabolism and promotes growth.
- Have students describe what happens after thyroxine enters a cell. *(It enters the nucleus and binds to a receptor. The hormone-receptor complex then binds to DNA, activating genes that produce enzymes.)*

Teaching Strategies

- Use this transparency when discussing the structure of a sperm cell and the anatomy of a testicle.
- Use the figure of the sperm to point out the two main structures of the cell. Tell students that sperm have been called genetic delivery machines.
- Use the figure of the testicle to show students where sperm cells are produced.
- Ask students to explain how the structure of a sperm cell helps it perform its function. *(Sperm cells consist of a head that is mostly nuclear material, mitochondria that provide ATP for energy, and a tail that propels the sperm and its genes toward an egg.)*

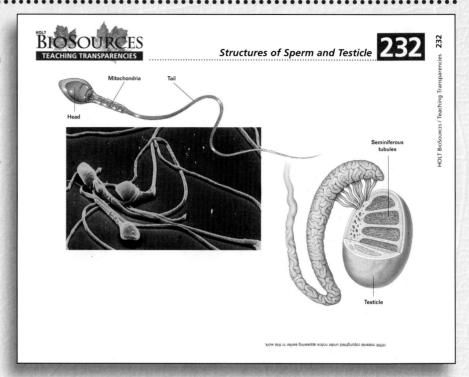

Teaching Strategies

- Use this transparency to illustrate the action of the drug Prozac® in a synapse.
- Have students compare the two sides of this figure. Explain that the right side of the figure shows the synapse before the drug is present. Note that the serotonin is being reabsorbed. Explain that the left side of this figure shows the synapse when the drug is present. Note that the serotonin is not being reabsorbed.
- Have students write a paragraph that describes how Prozac affects the levels of neurotransmitter present in a synapse. Be sure they explain how this affects the neuron and the individual taking the medication.

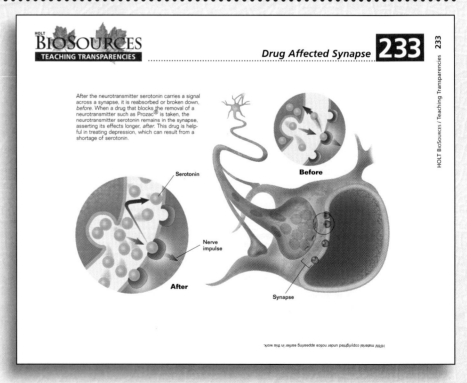

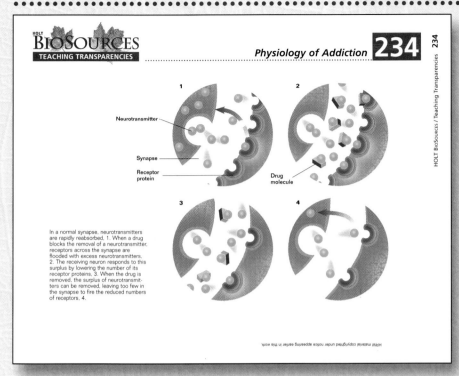

BIOSOURCES
TEACHING TRANSPARENCIES

Physiology of Addiction **234**

234

HOLT BioSources / Teaching Transparencies

Neurotransmitter

Synapse

Receptor
protein

Drug
molecule

In a normal synapse, neurotransmitters
are rapidly reabsorbed, 1. When a drug
blocks the removal of a neurotransmitter,
receptors across the synapse are
flooded with excess neurotransmitters,
2. The receiving neuron responds to this
surplus by lowering the number of its
receptor proteins, 3. When the drug is
removed, the surplus of neurotransmitters can be removed, leaving too few in
the synapse to fire the reduced numbers
of receptors, 4.

Teaching Strategies

- Use this transparency to illustrate an example of a mechanism that causes addiction. Review the terms *neurotransmitter, synapse,* and *receptor protein.*
- Guide students through the steps in this diagram. Point out the changes in the amounts of neurotransmitter in the synapse and in the number of receptor proteins.
- Have students explain why the neuron in part 3 has a reduced number of receptor proteins in its membrane *(because the drug molecules in part 2 have blocked the reabsorption of the neuro-transmitter, causing the cell to reduce its numbers of receptor proteins).*

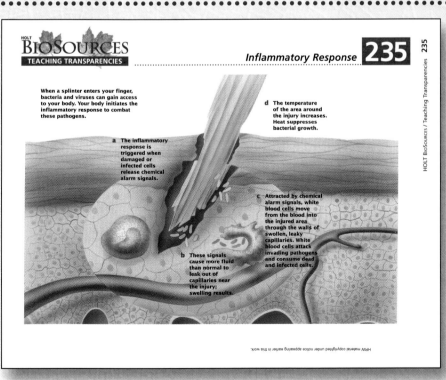

BIOSOURCES
TEACHING TRANSPARENCIES

Inflammatory Response **235**

235

HOLT BioSources / Teaching Transparencies

When a splinter enters your finger,
bacteria and viruses can gain access
to your body. Your body initiates the
inflammatory response to combat
these pathogens.

d The temperature
of the area around
the injury increases.
Heat suppresses
bacterial growth.

a The inflammatory
response is
triggered when
damaged or
infected cells
release chemical
alarm signals.

c Attracted by chemical
alarm signals, white
blood cells move
from the blood into
the injured area
through the walls of
swollen, leaky
capillaries. White
blood cells attack
invading pathogens
and consume dead
and infected cells.

b These signals
cause more fluid
than normal to
leak out of
capillaries near
the injury;
swelling results.

Teaching Strategies

- Use this transparency when discussing the events of the inflammatory response.
- Have students list the four classic symptoms of the inflammatory response *(heat, redness, pain, and swelling)*. Describe how the physiological changes mani-fested by each symptom help defend against pathogens. Remind students that increased temperature, though often considered an undesirable result of infec-tion, is beneficial because it inhibits bacterial growth.
- Have students explain what happens to blood vessels in the affected area. *(They swell, increasing the blood flow to the area.)*

Teaching Strategies

- Use this transparency to show how B cells and killer T cells are activated, thus triggering the immune response.

- Point out the central role of helper T cells. They stimulate the activities of killer T cells and B cells. Without helper T cells, neither type of cell can be activated. Emphasize that HIV, the virus that causes AIDS, destroys helper T cells and cripples the immune system.

- Have students describe the role of macrophages in the immune system. *(Macrophages stimulate helper T cells by consuming, processing, and presenting antigens to the T cells. This identifies the targets for destruction.)*

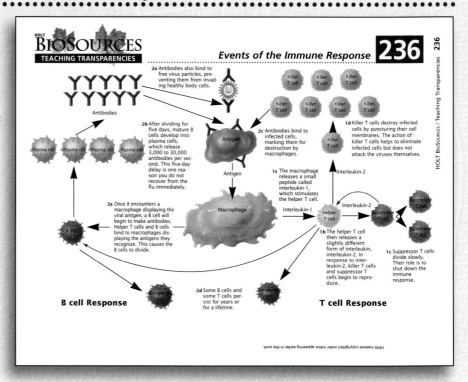

Teaching Strategies

- Use this transparency to illustrate how a killer T cell recognizes an infected cell.

- Inform students that all cells in the body have receptor proteins on their surface. Killer T cells have receptors that recognize the proteins of pathogens. These foreign proteins appear not only on the pathogens themselves but also on the membranes of infected cells, allowing killer T cells to recognize and destroy cells that have been invaded by the pathogen.

- Have students explain why it is advantageous for the body to destroy cells infected by a virus. *(Viruses turn the cells they infect into virus-making factories.)*

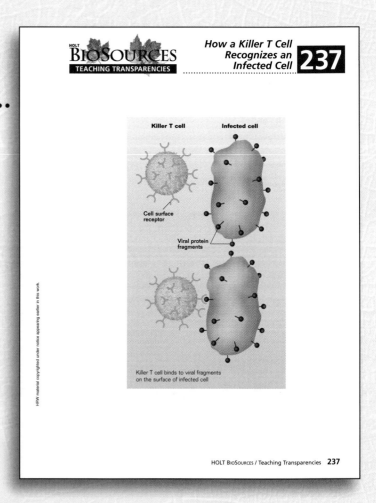

238

HOLT BioSources / Teaching Transparencies

- Vaginal, oral, or anal intercourse with an infected person
- Injecting drugs or other substances with hypodermic syringes or needles used by an infected individual
- Use of skin piercing equipment, such as tatooing needles, that has been used by an infected person
- From infected mother to fetus through the placenta; from infected mother to baby in breast milk
- Transfusions or injections of blood or blood products drawn from an infected person (Transmission no longer occurs by this route in the United States and other developed nations because blood is tested for the presence of HIV. It is still a transmission route in less developed countries, where such blood tests are often unavailable.)

Teaching Strategies

- Use this transparency when discussing the known routes of HIV transmission.
- Review the routes of HIV transmission listed in the table. Have students describe how they can reduce their risk of being exposed to the deadly virus HIV.